Economics

AND THE PUBLIC PURPOSE

John Kenneth Galbraith

Economics
AND THE PUBLIC PURPOSE

Houghton Mifflin Company Boston

Once again for C.A.G. with love

Acknowledgments

THERE ARE WRITERS, I am sure, who sit down with everything fully formed in their minds. I am not one, and the consequences are infinitely more taxing for those who help than for me. In the last three years parts of this book have gone through a dozen drafts and none fewer than four. And four faithful friends—Phyllis McCusker, now at the University of California, Mary Jo Hollender and Kirsten Voetmann, still blessedly of Cambridge, and my beloved ally Andrea Williams—have seen me through them all. I am grateful and truly beyond words. Hazel Denton checked my facts and footnotes, although if mistakes remain both the responsibility and the blame are mine. And William Raduchel helped me make these ideas an acceptable subject of instruction. I thank both of these colleagues. I thank also the students at both the Cambridges— that on the Charles and the one in England—who, with varying patience, heard me out on these matters in the last three years and who, I trust, suffered no lasting damage from versions I subsequently abandoned.

Foreword

THIS BOOK is in descent, the last in the line, from two earlier volumes—*The Affluent Society* and *The New Industrial State*. There are also some genes, though not many, from yet another volume—*American Capitalism*. The principal precursors each dealt with a part of the economic system; this one seeks to put it all together, to give the whole system. The earlier volumes were centrally concerned with the world of the great corporations—with the decisive part of the economy which the established or neoclassical economics has never ingested. There is also the world of the farmer, repairman, retailer, small manufacturer, plumber, television repairman, service station operator, medical practitioner, artist, actress, photographer and pornographer. Together these businesses supply about half of all we use or consume. This book seeks to bring them fully into the scene. In economics as in anatomy the whole is much more than the sum of the parts. This is certainly so when the parts are in support of each other or in conflict with each other or are otherwise shaped by the fact of their common existence. Also, a lesser point, the earlier books stopped at the water's edge. This one gives the first elements of the international system.

The New Industrial State pictures the world of the large corporation as the outgrowth of the neoclassical world of monopoly and oligopoly. At least by implication what was left behind was the world of the competitive market. That also I here correct; what is left behind is, in fact, something resembling the neoclassical admixture of entrepreneurial mo-

nopoly, oligopoly and competition. The consequence of so
seeing matters is a better explanation of the behavior of the
entrepreneurial firm and what I here call the market system.
It shows, also, that the world of the large corporation is something new—that it is a clear break with what is described by
traditional doctrine.

The traditional economics assumes that economic institutions and the motivation of the people who comprise them
change but slowly. As with physics or botany economic truth
once established is largely immutable. This is agreeable but
not so. Economic institutions change rather rapidly; the large
corporation and its relations with the community and state are
especially in flux. And with such change comes new information, new insight. In consequence the rate of obsolescence in
economic knowledge is high. On many smaller matters such
change and such information have nurtured a view that differs
from that of the earlier volumes. And I must remind the
reader, and more painfully myself, that on the present views
time will also have its way.

There is also the disconcerting business, half euphoric, half
depressant, of discovering what you did not see before. One
such matter is central. Dominating this book as a drumbeat
is the theme of unequal development and the associated inequality in income. The unequal development is unrelated to
need; the inequality in income bears no necessary relation to
productivity or efficiency. Both are the result of unequal deployment of power. Both are socially damaging. In the established economics these tendencies are mostly concealed and
where not concealed are misconstrued. But I did not see them
with sufficient clarity in the earlier books. In *The Affluent
Society* I dealt with the starvation of the public services as
though all services were alike. I did not see that this deprivation was great where public needs were involved, nonexistent
where powerful industry pressed its requirements on the state.
And perhaps partly because I was dealing only with the world
of the large corporation, I did not identify adequately the

systemic inequality in product and income as between different parts of the so-called private economy. Nor did I
identify sufficiently the problem, unknown to orthodox economics but endemic to planning, of matching performance in
related parts of the economy and the consequences of failure
to do so. From this failure come blackouts and energy crises
of which we will hear much more before we hear less.

2

This book also speaks in a somewhat different mood to a
somewhat different audience. In recent years the ideas in the
earlier books have won a certain measure of professional acceptance, especially among a younger generation of economists. It would be nice to think that this was the result of the
logical force and clarity of the argument; more of my debt,
alas, is to circumstance. In the earlier books I argued that the
quality of our life would suffer from a single-minded concentration on the production of salable goods as a social goal;
that the environment would be a casualty; that we would suffer
especially from the disparate development of the services of
the private economy and those of the state; and that some exceedingly somber problems were inherent in the growth of the
power of the great private and public bureaucracies and their
exercise of that power, including that over weaponry and
other technical development, in their own interest. This bureaucratic power, not that classically associated with the
sovereignty of the consumer, was now the decisive force in
economic and political life.

The initial reaction to this argument, at a time when the
prospect seemed more benign than now, was less than enthusi-
astic. Important economists, including many whose self-regard
is widely shared, were sharply averse. They judged the traditional ideas to have survived intact. But events intervened,
and with a force which I did not foresee. Problems inherent
in my case—the decay of services and therewith of life in the

great cities, pollution and environmental disharmony, the extravagances and dangers of the weapons culture, the seeming indifference to the public will of the great corporations—became ever more visible and the staple subjects of oratory and, on occasion, even of action. For being right, one may perhaps conclude, it is better to have the support of events than of the higher scholarship.

The notion that power in the modern economy lies increasingly with the great organizations and increasingly less with the supposedly sovereign consumer and citizen has also been making its way into the textbooks. Something here is owing to a vacuum. In recent years there has been a rapidly growing discontent with the established or neoclassical model of economic and political life. The way was open for an alternative. Still the inertial forces are great. The textbook writer is naturally a cautious fellow. Like liberal candidates for public office he must always have one eye for what is reputable and salable as distinct from what is true. And, as these pages will sufficiently emphasize, economics is not primarily an expository science; it also serves the controlling economic interest. It cultivates the beliefs and therewith the behavior that such interest requires. I would like to see economic instruction reflect the current reality. And it should be no part of its purpose to propagate the convenient belief. In writing this book, I have had the general reader in view. But I've also had the emancipation of the student from the textbook very much in mind.

Not that one can escape entirely from the textbook. Economics now brings its communicants to conclusions that are convenient for the great corporation but painful for the society. But the economic concepts and institutions that are explained by the textbooks—capital, rent, Gross National Product, index numbers, money supply, income tax, the capital market, the purposes of the Federal Reserve System—are essential knowledge. So is the capacity to visualize an economic system, provided always that one does not become a captive of a particular image. So we must still have textbooks—though hopefully we will not always have their present view of the economic system. And we must not be limited to the textbooks.

Perhaps it should be added that we must also still have diligence. In recent times the politically emancipated, or those who so regard themselves, have tended to identify difficult matters with the obscurantism of the Establishment. Study is a tedious disguise for wickedness, a way of diverting people from the simple disconcerting truth. This does not arouse my sympathy. It would not, the more radical may be reminded, have aroused the sympathy of Marx, for his was a notably demanding intellectual tradition. It is one thing to liberate man from physical toil. To exempt him from mental effort is premature.

3

As noted, this book seeks to bring the market sector of the economy and therewith the whole economic system into focus. But if we are to see the whole, the highly organized sector with which I dealt in *The New Industrial State* must also be here. In consequence some chapters go over ground covered in the earlier book. Such repetition rightly arouses suspicion—few things are more tempting to a writer than to repeat, admiringly, what he has said before. Yet there was no alternative. Pleas to the reader to go back to an earlier book are poorly obeyed. However, I've greatly abbreviated the previous argument and, as noted, also altered the tone. Before, I was seeking to establish a bridgehead in existing belief. This, unless I am greatly mistaken, has now, at least partly, been won. Accordingly, where the earlier book argued (and in some degree shamed or cajoled), this one assumes a receptive audience and explains.

Once I considered publishing the part of this book dealing with the theory of reform as a separate volume. (For better or for worse I am a reformer and not a revolutionist.) It is a terrible and sobering fact that the first part of any book, and especially one on economics, is likely to be better read than the last. But I could not have two books without prefacing the second with a detailed and repetitious summary of the first. This would have been a heavy trial for the reader and a dubious enjoyment for the writer. (I recapitulate at intervals in this volume not for pleasure but because such reminder is recurrently necessary if the further argument is to be persuasive.) Had there been two books, moreover, I would have had to end the first with a request to the reader to buy another. Again one doubts the response. So there is only one book. But I beg the reader not to give up after the first twenty or so chapters. It is then that the book gets on to the questions of what to do. By these I set much store. For on no conclusion is this book more clear: Left to themselves, economic forces do not work out for the best except perhaps for the powerful.

Contents

ONE

THE FOREST

The Uses of an Economic System
—and of Economics

THE PURPOSE of an economic system would seem, at first glance, to be reasonably evident, and it is commonly so regarded. Its purpose is to provide the goods and render the services that people want. In the absence of such a system—one that grows food, processes, packages and distributes it, manufactures cloth and makes clothing, constructs houses, furnishes them, supplies educational and medical services, provides law and order, arranges the common defense—life would be difficult. Thus the function. The best economic system is the one that supplies the most of what people most want.

Though greatly celebrated in the textbooks this is rather too simple a view. Over the last hundred years numerous economic tasks have come to be performed by organizations—by industrial corporations, electric utilities, airlines, merchandising chains, banks, television networks, public bureaucracies. Some of these organizations are very large; as few would doubt, they have power, which is to say they can command the efforts of individuals and the state. They command these, most will agree, for their own purposes, these being the purposes of those who participate through membership or ownership in the enterprise. Perhaps by some miracle of accident or design these purposes are usually the same as those of the public. In the absence of such miracle or arrangement it is, not surprisingly, the purposes of organization, not those of the public, which are served.

So viewed, the function of the economic system is no longer simple—at least for anyone wishing to see the reality of things. Partly the economic system serves the individual. But partly it is now seen to serve the needs of its own organizations. General Motors exists to serve the public. But General Motors also serves itself as well or instead. Not many will find such a proposition radically in conflict with common sense. To quite a few it will seem trite. It is only remarkable in being at odds with the main thrust of economics as it is traditionally taught. A shrewder view does, in fact, accept what is trite. It seeks to identify the interests the great organizations pursue, how they conduct the pursuit and with what effect on the public.

2

With a revised view of the purpose of the economic system goes a revised view of the purpose of economics. So long as the economic system is imagined to be in the ultimate service of the individual—to be subordinate to his needs and wishes— it can be supposed that the function of economics is to explain the process by which the individual is served. Economists, like other scholars, cherish definitions of their subject which convey a sense of profound and universal meaning. The most famous of these states economics to be "the science which studies human behavior as a relationship between ends and scarce means which have alternative uses."[1] The most influential teacher of our time puts it a trifle more simply: "How . . . we choose to use scarce productive resources with alternative uses, to meet prescribed ends . . ."[2]

These definitions will seem admirably forthright. People are

[1] Lionel Robbins, *An Essay on the Nature and Significance of Economic Science,* 2d ed. (London: Macmillan, 1935), p. 16.
[2] Paul A. Samuelson, *Economics,* 8th ed. (New York: McGraw-Hill, 1970), p. 13.

making decisions on what they will have; firms are deciding how these decisions can best be served. Economics studies the behavior of people so engaged. It is a science because it has no purpose except to understand that behavior.

But if it is assumed that the organizations that participate in this process have power—that it is their purposes that are served and that people are bent to these purposes—even the minimally alert must ask: Is it not possible that economics also serves the purposes of organization? Organizations have power. Will they be without influence on the subject which deals with them and that exercise of power? Could the definitions just offered be a cover for that power?

People are right to ask. Economics provides them with their image of economic society. That image notably affects their behavior—and how they regard the organizations that comprise the economic system. If the image provided by economics makes goods or the capital, labor and materials by which they are produced scarce, it is because goods are important— the fulcrum on which well-being and happiness depend. The process by which goods are made becomes, thus, a matter of prime social urgency. Great importance will be attached to the organizations that produce goods, and much prestige will accrue to those who manage and lead these organizations. The burden of proof will be heavily on any action—any regulatory action by the government, any tax, any work rule of a union— that seems to interfere with production or which those concerned say will so interfere.

The imagery of choice has a yet more important effect. It means that this choosing—the decision to purchase this product, reject that—is what, when aggregated, controls the economic system. And if choice by the public is the source of power, the organizations that comprise the economic system cannot have power. They are merely instruments in the ultimate service of that choice. Perhaps the oldest and certainly the wisest strategy for the exercise of power is to deny that it is possessed. Monarchs, including the most inimical of despots, long pictured themselves as the mere projection of divine will. This the established religion then affirmed. It followed that their behavior, however scandalous, expensive and damaging to health, life, livelihood or common decency, could not be questioned, at least by the true believer. It was in the service of higher will. The modern politician perpetuates the same instinct when he explains, however unconvincingly, that he is only the instrument of his constituents, the expression not of his own preferences but of the public good.

Though avowedly more secular, economics and particularly the imagery of choice in the market puts the business firm similarly in the service of a higher deity. In consequence it is not responsible—or is only minimally responsible—for what it does. It responds to the theistic instruction of the market. If the goods that it produces or the services that it renders are frivolous or lethal or do damage to air, water, landscape or the tranquility of life, the firm is not to blame. This reflects the public choice. If people are abused, it is because they choose self-abuse. If economic behavior seems on occasion insane, it is because people are insane. Socially objectionable exercise of power by organizations in their own interest is thus exorcised or largely exorcised from formal economic thought. Should there be, in fact, such exercise of power, it will be seen how convenient is this contrary belief—and how worthwhile its cultivation.

Cultivation of useful belief is particularly important because of the way power is exercised in the modern economic system. It consists, as noted, in inducing the individual to abandon the goals he would normally pursue and accept those of another person or organization. There are several ways of accomplishing this. The threat of physical suffering—prison, the lash, an electrical impulse through the testicles—is in an ancient tradition. So is economic deprivation—hunger or the disesteem of poverty if one does not work for the wages and therewith accept the goals of an employer. But persuasion— the altering of the individual's belief so that he comes to agree that the goals of another person or organization are superior to his own—is of increasing importance. That is because in modern society physical force, though still applauded by many in principle, attracts adverse comment in practice. And with increasing income people become less vulnerable to the threat of economic deprivation. Accordingly persuasion (in forms later to be examined) becomes the basic instrument for the exercise of power. For this the existence of an image of economic life that is congenial to the organizations that are exercising power is vital. So is instruction that implants this image. It persuades people that the goals of organization are really their own or paves the way for such persuasion. An image of economic life which makes people the instruments of organization goals would be far less useful or convenient.

The contribution of economics to the exercise of power may be called its instrumental function—instrumental in that it serves not the understanding or improvement of the economic system but the goals of those who have power in the system. Part of this service consists in instructing several hundred thousand students each year. Although gravely inefficient this instruction implants an imprecise but still serviceable set of ideas in the minds of many and perhaps most of those who are exposed to it. They are led to accept what they might otherwise criticize; critical inclinations which might be brought to bear on economic life are diverted to other and more benign fields. And there is great immediate effect on those who presume to guide and speak on economic matters. Although the accepted image of economic society is not the reality, it is what is available. As such it serves as a surrogate for the reality for legislators, civil servants, journalists, television commentators, professional prophets—all, indeed, who must speak, write or act on economic questions. It helps determine their reaction to the economic system; it helps set the norms of behavior or action—in work, consumption, saving, taxation, regulation—which they find good or bad. For all whose interest is protected thereby this is a very useful thing. My present effort will be thought by many to be less useful. It eschews the instrumental function of economics. It reverts to the older, more traditional, more scientific, expository purpose which is to seek to understand how things are.

3

In considering the sources of the instrumental role of economics, nothing should be attributed to conspiracy and not much to design. Economists are not deliberately subservient to economic interest. Few strive consciously to conform. Rather the dominant economic interest is the standard and accepted voice in the community. What it approves and finds convenient is sound policy. What it disapproves or finds inconvenient may be interesting or imaginative, but it is not a worthy guide to responsible belief or action. Economists, like other people, have an instinct for what is worthy, responsible and reputable. It is by defining what is responsible and reputable that economic interest principally prevails.

Much of the instrumental service of economics is a byproduct of its history. It happens that the image of an earlier economic society serves admirably the instrumental purposes of a later one. It has been necessary only to cling to and affirm as immutable the earlier truth. Economics as a discipline took form when business enterprises were small and simple and agriculture engaged most of the productive energy of people. Firms responded to changing costs of production and to changing market prices. They were subordinate to the instruction of the market. The theory reflected the fact. In time the theory was amended to embrace monopoly—or, more precisely, oligopoly—but it remained the captive of its origins. The competitive firm was still the centerpiece. And the oligopolist also responded to market movements and was impelled to do so, for he sought single-mindedly to maximize profits. Thus the market and hence the consumer remained sovereign. Consumer choice continued to control all. Economics thus slipped imperceptibly into its role as the cloak over corporate power.[3] No one managed the process. What seemed a decent intellectual conservatism became powerful support to economic interest. To all this I will return.

The instrumental role of economics is not, however, immutable. In recent times, and especially among younger economists, there has been a sharp revolt. The image that has so long dominated instruction, shaped theoretical speculation and disguised the exercise of economic, social and political influence by producing organizations has come under general attack. What was once economics and accepted is now neoclassical economics, and the term is strongly suggestive of obsolescence.

There have been several causes of this rebellion. In part the power which this model protects has become too palpable; it is no longer intellectually decent to try to disguise it.

Also there are the increasingly formidable consequences of the exercise of power. People can be persuaded and scholars can persuade themselves that General Dynamics or General Motors is responding to the public will so long as the exercise of its power does not threaten public existence. When ability to survive the resulting arms competition or breathe the resulting air is in doubt, persuasion is less successful. Similarly when houses and health care are unavailable and male deodorants are abundant, the notion of a benign response to public wants begins to buckle under the strain.

The changing character of the universities has also been a factor. They have increased greatly in size and complexity in recent decades in response to industrial need. And in consequence of their size and importance—a fortunate paradox—

[3] As I have argued elsewhere, cf. *The New Industrial State*, 2d ed., rev. (Boston: Houghton Mifflin, 1971), pp. 403–406, specialization within the subject matter of economics keeps scholars from any need to reflect on the larger truth or role of the subject. The good scholar is the man who sticks tightly to his own last, declines any concern with the truth or error of the system of which his work is a part. And such concern, since it involves the difficult task of offering more satisfactory alternatives, can usually be attacked as deficient in methodology or proof.

they have become an increasingly independent force in themselves. Once the dissident faced sanctions, he might be found out and fired. Now he merely foregoes the reputable applause. This a man of modest courage can face. There is little doubt that revolutions tend to break out in the United States at the point in history where they have become comparatively safe. So with this one.

4

Organization develops very unevenly in the economic system. It reaches its greatest scale in communications and the automobile industry, its greatest technical complexity and most intimate relation to the state in the manufacture of weapons. In agriculture, housing construction, the service industries, the arts, the more uncomplicated forms of vice, the business firm remains relatively simple. With these differences go very great differences in power and consequent social effect. Ford, Shell and Procter & Gamble deploy much power. The individual farmer has no such power; the residential builder has very little. These differences, in turn, have much to do with how the economic system performs—and for whom. Here, rather more than in the original eccentricities of consumer or citizen taste, is the explanation for the high level of automobile, highway and weapons development, the low level of development in housing, health and nutrition.

It follows that the economic system cannot usefully be dealt with as a single unit. Ideally it should be considered as a continuum—a procession of organizations extending in the United States from the simplest surviving family farm at the one extreme to American Telephone and Telegraph and General Motors at the other, and similarly from peasant to Volkswagen in other industrial countries. But classification, even though it involves arbitrary lines, is the first step toward clarity. Little is lost and much is clarified by dividing business organization between two classes, those that deploy the full range of the instruments of power—over prices, costs, suppliers, consumers, the community and the government—and those that do not. The fourth chapter hence outlines this two-part model of the modern economy. But first it is necessary to have the major lineaments of the established or neoclassical system clearly in view. We must know the bonds we are seeking to break.

The Neoclassical Model

The purpose of studying economics is not to acquire a set of ready-made answers to economic questions, but to learn how to avoid being deceived by economists.

—Joan Robinson

THE ACCEPTED INTERPRETATION of the nonsocialist economic system is called the neoclassical model by economists, and economics by others. Its principal origin is in Adam Smith's *Wealth of Nations,* published in 1776. The 200th anniversary of this superlative book will coincide with that of the birth of the American Republic; the 250th anniversary of Smith's birth has just been celebrated by the Royal Burgh of Kirkcaldy. Smith's ideas were greatly developed in the first half of the last century in Britain by David Ricardo, Thomas Malthus and James and, more particularly, John Stuart Mill, and came to be denoted the classical system. In the last quarter of the last century the addition of so-called marginal analysis by Austrian, British and American economists led eventually to the substitution of the term *neoclassical* for *classical* economics. In the nineteen-thirties two further and major amendments were made: Until then markets were assumed to be served by numerous firms, each supplying a small share of the total product. All were subject to a market price that none controlled. There were monopolies, but they were an aberrant exception. Now it came to be accepted that numerous markets might be dominated by a few large firms which exercised collectively the power hitherto associated with one. This was oligopoly. And with the publication and wide acceptance of

Keynes's *General Theory*[1] the system was no longer assumed to be self-regulating. Only active intervention by the state would keep the economy at or near full employment and ensure its steady growth.

Additionally, in the last forty years, the neoclassical system has undergone much refinement. Indeed so diverse and specialized is this detail that no individual economist pretends to a knowledge of more than a fraction of the total. To a substantial degree the neoclassical system now exists for the refinement that it sustains—this has become an end in itself. But the refinement does not affect or even touch the central substance of the discipline. That, however subjectively, is deemed to be in final form.

2

The essence of the neoclassical system is that individuals using income derived, in the main, from their own productive activities express their desires by the way they distribute this income for the various goods and services available to them in markets. Their tendency, deriving from the marginal analysis just mentioned, is so to distribute their income that the satisfaction derived from the last unit of expenditure for any particular purpose is equal to that from the expenditure for any other purpose. At this point satisfaction, even happiness, is maximized. No judgment is passed upon the desires of the individual; their source is not much examined. Though doubtless shaped by the culture they are an expression of individual personality and will. There they begin. That is sufficient.

The foregoing expression of the individual's will is passed on by the market to the producer along with the similar expression of others. Where the desire is strong, so will be the willingness to spend money. And so will be the price in the market. Where the desire is weak, so will be the price. The

[1] John Maynard Keynes, *The General Theory of Employment Interest and Money* (New York: Harcourt, Brace, 1936).

producer is motivated, for the purposes of the neoclassical model exclusively, by the prospect for profit. This, over an unspecified period of time, he seeks to maximize. Price changes signal to this motive. Included among the recipients of the information so transmitted are producers who can expand or contract their production, others who can enter or depart the business. They respond. In such response they ensure that production is ultimately at the command of the individual.

Information also passes from the producer to the market and to the consumer. This, however, involves no similar command; rather it is intelligence on the basis of which the individual or consumer alters his instructions to the producer. Specifically, if there is a change in the technical conditions of production which reduces costs, the margin of profit of the producing firm will be increased. Producers, new and old, will respond to this opportunity with increased production, and, in consequence, prices will fall. This is advice to the individual—the consumer—that he should reconsider the distribution of his expenditure to accord with the new opportunity for enhancing his enjoyment. He does so and thus further informs the producer as to his will.

In formal theory the fact that instruction originates with the consumer is not greatly emphasized. The view is of an apparatus by which information is passed from consumer to producer and from producer to consumer. No judgment is passed on this machinery; one does not ask whether a typewriter or a corn harvester is good or bad, only how and how well it works. However, the moral sanction of the system depends profoundly on the source of the instruction. This comes from the individual. Thus the economic system places the individual—the consumer—in ultimate command of itself. This economic theory is associated with a political theory which places the citizen, as a voter, in ultimate authority over the production of public goods[2]—over the decision to have more

[2] Although the theory is far less fully developed. In consumer markets the power of the individual is expressed in formal models. Where the state is con-

expenditures for education or for weapons or for space travel. These economic and political theories are basic to a larger image of a democratic (or at least nonauthoritarian) society which is comprehensively subordinate to the ultimate power of the individual. The individual being in charge, he cannot be in conflict with the economic or political system. He cannot be in conflict with what he commands.

3

The control of the economic system by the individual—by the consumer or citizen—does not mean that power is distributed equally. It is a well-established point that the citizen who votes ten times in an election has, all else equal, ten times more power than the citizen who votes but once. Likewise in the more common case of the man who controls ten votes as compared with the man who has only his own. And similarly the man who spends $70,000 in the course of a year speaks to the market with ten times as much authority on what is produced as does the man who disposes of but $7000. This, in the democratic imagery, is a flaw. But power still rests with the individual. It is only that, in the exercise of that power, some individuals are more equal than others.

Also in the original or classical version of the neoclassical model the differences in power associated with differences in spendable income tended to correct themselves. Producers being numerous in each market (the exceptional case of monopoly apart), none had power to influence the common price. As with the modern wheat or cattle grower, to charge more than the market price was to sacrifice all witting customers. Who would pay more than the going price if he could

---

cerned, it remains a matter of rhetoric and subjective belief. Since the eighteenth century, "economic theory based on personal utility calculations has prospered while the political theory based on these same calculations has languished . . ." Edwin T. Haefele, "A Utility Theory of Representative Government," *The American Economic Review,* Vol. 61, No. 3, Pt. 1 (June 1971), p. 350.

get all he wanted at the going price? To charge less than the market price—since all could be sold at that price—was merely insane. Why take less when one could have more? Since each producer's output was very small in relation to the total, none calculated the effect of his production and sales on the market. If prices and profits and therefore income were exceptionally favorable, some or all producers would be induced to expand production. Some others, as noted, would be induced to enter the business, and since it was assumed that the firms were generally small and the required capital of manageable size, this would be practical. The expansion in output would lower the market price—the price that none controlled—and therewith the resulting profit and income. This, in turn, would reduce the power, i.e., the purchasing power which the producer deployed as a consumer. Thus not only was the consumer ultimately in control, but built into the system was a powerful force, that of competition, which acted to limit or equalize income and thus to democratize that control.

4

By the nineteen-thirties the assumption of competition—of many firms, by necessity small, participating in each market —had become untenable. Since late in the previous century the giant corporation had become an increasingly obtrusive feature of the business landscape. Its importance was assumed everywhere except in the economics textbooks. And even the more casual scholars had difficulty in disguising from themselves the fact that markets for steel, automobiles, rubber products, chemicals, aluminum, other nonferrous metals, electrical gear and appliances, farm machinery, most processed foods, soap, tobacco, intoxicants and other basic products were shared not by many producers, each without power over its prices, but by a handful of producers with a great deal of such power. Accordingly the neoclassical model was modified to embrace the case of markets shared by two, three, four or a handful of (usually) very large producers. Between the competition of the many and the monopoly of the single firm there was now inserted the oligopoly of the few. And, although at first reluctantly, oligopoly came to be recognized as a normal form of market organization.[3]

However less was altered by this amendment than was then or is now imagined. The structure of the business firm was not thought to have changed. Nor was its motivation. "The firm is the primitive concept of the [accepted] theory. It is assumed implicitly or, on occasion explicitly, that the firm is run by an individual owner who is a profit maximizer."[4] What was added was the power to set the price for its product or service. The price so established by one firm affects the price that can be set for the same item by any other. Recognizing this, each firm gives thought to the common interest of the industry—to the price that is acceptable to all. This is the price, motivation being unchanged, that will maximize return. Thus, in the neoclassical model, an oligopoly is assumed (though, because of inadequate information or some unnecessary sales costs, perhaps imperfectly) to achieve the same result as a monopoly. Instead of one firm's maximizing its revenues from the sale of the product, a small number maximize the revenues they share. All firms—a vital point—remain at the command of the consumer. The message of the consumer in the form of increased or diminished purchases is still transmitted to the market; this is still the instruction, the only instruction, to which the firm and the industry respond. This instruction tells them where they can find the greatest possible profit, which is their sole interest. So the consumer is still in control.

[3] Although something was owing to the work of antecedent writers, notably that of Piero Sraffa of the University of Cambridge, the decisive books were the late Edward H. Chamberlin's *The Theory of Monopolistic Competition* (Cambridge: Harvard University Press, 1933) and Joan Robinson's *The Economics of Imperfect Competition* (London: Macmillan, 1933).
[4] Martin Shubik, "A Curmudgeon's Guide to Microeconomics," *The Journal of Economic Literature,* Vol. 8, No. 2 (June 1970), p. 411.

The profits of the oligopoly are higher than they need be, and the competitive process by which these, and the resulting revenue, are brought back to normal is impaired. So the system no longer has the same tendency to equality in income as before. And since prices are higher than they need be, production and therewith investment and employment where there is oligopoly (or monopoly) are smaller than would be ideal.[5] But the ultimate social, political and moral sanction of the system deriving from its ultimate subordination to the will of the individual remains unimpaired.[6]

The inequality resulting from monopoly or oligopoly is also confined to comparatively few people—and thus, in principle, can be remedied by public action. Workers do not share in the monopoly gains, for the monopolist has no inducement to pay above the going rates for labor. If wages were, in fact, higher in a monopolistic sector, the influx of workers would soon bring them back down. So only the proprietor would get an undue return. A sufficient income tax could take care of him.

It should be noted that exponents of the neoclassical system, while they have long deplored the monopolistic and hence pathological tendencies of oligopoly in principle, have

[5] Not smaller than they would be under competition, as the textbooks often say. Were there competition, there would be many firms, each much smaller and with a different technology and different cost functions. One does not know, accordingly, whether the competitive equilibrium would be at a greater or smaller level of output, investment or employment. The monopolistic result cannot, accordingly, be compared with that under competition. The comparison is much made by economists and is licensed, as it were, by the seeming moral superiority of competition.

[6] However the notion of the competitive market also has a high survival value. ". . . the economist sees the competitive market and its pricing mechanism as a particularly efficient way of giving expression to individual choices . . . Free choice and competition expressed through purchasing and selling decisions of individual competitors often have a remarkable property of yielding social results that cannot be improved on by public action. At the same time, the market has many defects and shortcomings that require public action. In dealing with these defects, the economist is happiest when he can recommend a public policy that works to perfect the market rather than to overrule or finesse it." Arthur M. Okun, *The Political Economy of Prosperity* (Washington: The Brookings Institution, 1970), pp. 5–6.

never done much about them in practice. There was cancer, but one did not operate. Before the notion of monopoly was enlarged to include oligopoly in the nineteen-thirties, the identifiable case of monopoly was rare. In large-scale industry only the Aluminum Company of America came reliably to mind. Until then talk of regulating, socializing or breaking up monopoly was not wholly unrealistic. But once oligopoly came to be recognized as a dominant market form, such remedy became tantamount to talk of socializing, regulating or breaking up the firms that composed the dominant part of the economic system. This was not remedy but revolution. Economists are not revolutionaries nor are their textbooks. Also, although this was not recognized with large-scale organization, there came a change in the motivation of the firm, a point soon to be examined. Insufficient resource use gave way to relative overuse. Economists continued to pay lip service to the antitrust laws; the laws continued to enjoy a deeply impassioned defense from the lawyers they supported. But the most important service of both the economists and the lawyers, as we shall presently see, was to siphon critical attitudes toward the large corporation into an alley that was safely blind.

The firm in the neoclassical model is also assumed to be fully subordinate to the state. The economic management of the state is responsive to the needs of the public as a whole and not of the business firm. The numerous services (supply of educated manpower, support to technology, highways for cars) that industry requires are provided in response to the larger public interest. The same is true of products and services that are sold to the state; these—weapons, weapons development, other research and development, support to space exploration—are a reflection of legislative and, ultimately, of citizen choice. The citizen is the ultimate arbiter of both the total volume and the particular kinds of public services. To these matters we now turn.

The Neoclassical Model II:
The State

IDENTIFIED WITH THE NEOCLASSICAL MODEL is the neoclassical state. The economic system functions in response to the instruction of the market and ultimately of the consumer. Where, for one reason or another, the response to this instruction is inadequate or imperfect, the government may be required to amend the instruction or supplement the response so that it accords better with the public interest. Firms respond well to market and consumer instruction on behalf of heroin, massage and carcinogenic agents. Such response is not thought socially desirable, moral or healthy; here, accordingly, the instruction of the market is superseded. There are other services—the common defense, education, protection of people and property—which the market does not reliably supply. Here too the state must act.

However there is a strong neoclassical presumption that most economic tasks will be accomplished in response to the instruction of the market. The state is supplementary and regulatory in its role, and the burden of proof is implicitly on those who say that its action is needed. At least until comparatively recent times it was assumed that the tasks performed by the state, as compared with those accomplished by private firms in response to market instruction, would be decidedly secondary in scale.

Unfortunate and exceptional aberration apart, the neoclassical state is superior to economic interest and superior, in particular, to the influence or power of the business firm. The latter is subordinate to the market and thus to the consumer; being so in leash, it cannot be a dominant force in relation to the state. And the state, being subject to the instruction of the citizen and voter, cannot be beholden to other power.

There is, however, aberration. The business firm seeks, naturally enough, to influence the market by which, otherwise, it is enslaved. It may seek tariffs that will exclude supply and thus raise prices in its markets. Or it will want its prices supported by government purchases. Or it may seek to have the government bar innovation that threatens to provide a better product at the same price or the same one at a lower price. Or it may seek the support or acquiescence of the government in devouring its competitors and thus winning control of its prices. Neoclassical economics has a strongly adverse attitude toward tariffs, price supports, suppression of technological innovation and anything that suggests government assistance to, or acquiescence in, monopoly. These, all involving the rigging of the market in favor of the firm, are the classical devices for winning public support for private purpose.

The neoclassical model is also faithful to its origins in what it ignores. The state, as just noted, was deemed to have a small, relatively subsidiary role in the economy as a whole. It was not a large purchaser of products; the services of the state, though of general importance for economic progress, were not—in the manner of support to research and development or the provision of highways for automobiles—decisive for the development of particular industries. The influence of the private firm on the government as regards procurement of its products or provision of needed services, having no deep historic roots, was not much considered. Nor is it now.

2

Within the last half century the neoclassical view of the state has been amended to include among the state's functions the need to provide overall management for the economy. This management also is seen as being superior to particular economic interest. It too reflects the general public interest.

Until the Great Depression of the nineteen-thirties, a decade that left a deep imprint on all branches of economic thought, the economic system was regarded in all neoclassical attitudes as being self-regulating—more precisely, self-righting. There might be temporary malfunction, but its basic tendency was to employ all willing and available workers for something close to maximum output. This was because production provided the income which purchased that production and, in the end, provided enough to purchase it all.

To be sure, some of the income so provided might be saved —not spent. But what was saved would, in the end, be invested—which is to say, it would also be spent. Were savings temporarily excessive, interest rates would fall, and this would encourage their use. Or, demand being momentarily deficient, prices would fall, and a smaller supply of purchasing power would suffice to clear markets. So there could be no permanent shortage of purchasing power. By the nineteen-thirties the notion that production created its own sufficient demand had been economic scripture for more than a century. It was given formal expression as Say's Law of Markets. Whether or not a person accepted Say's Law was, until the thirties, the prime test by which economists were distinguished from crackpots.

The neoclassical equilibrium between production and the purchasing power that acquired the resulting product was established at the level where all willing and useful workers were employed. If men were unemployed, wages would fall in consequence of the competition for jobs. It would be profitable to hire more men. Perhaps, in consequence of the lower wages, demand and prices would fall. But wages—directly influenced by the unemployment—would fall more. This, the fall in real wages, would be the decisive factor in expanding employment. The expansion would continue until all were employed.

It was the historic achievement of John Maynard (later

Lord) Keynes in the mid-thirties, in the wake of numerous less reputable voices (who lacked Keynes's prestige and the savage reinforcement of the circumstances of the Great Depression), to destroy Say's Law virtually without trace and therewith the illusion of a self-righting economy. After Keynes it was accepted that there could be a shortage (or surplus) of purchasing power in the economy and that neither wages nor interest rates reacted usefully to correct it. A reduction in wages might merely reduce purchasing power—aggregate demand, as it came to be called—yet more and make things worse. In the absence of sufficient demand even the lowest interest rates, as the Depression experience showed, would not encourage the needed investment and thus enhance demand. Stagnation would continue. The only answer was for the state to intervene.

The state could spend in excess of its tax revenues and so add to demand when this was required. And if demand were in excess of the current capacity of labor force and plant—so forcing prices up—it could reverse the process. This was to bring fiscal policy to the support and control of the economic system.

Additionally the government could manipulate the supply of funds available for lending and, therewith, the interest rate at which such funds were available. Alone, low interest rates might not accomplish much. As part of a general strategy of stabilization, monetary policy would be effective. Although economists have derived much useful prestige from the mystery which is supposed to surround monetary policy, in essentials it is rather simple. Savings deposited with the banks or other financial institutions are, of course, available for relending. The amount that is so available can be extended by allowing the banks to borrow from the central bank—in the United States, the Federal Reserve System. This can be encouraged, as necessary, by a favorable lending (rediscount) rate. By buying government securities from the banks, thus leaving them with the money, the supply of funds which they have for lending can be further enhanced by the central bank. If the need is for contraction of demand, the process can be reversed. We may note that an active monetary policy makes the interest rate a planned or fixed price—the fixing being done by the central bank. This point is not, however, emphasized in the neoclassical model. Whether the interest rate is a market- or a publicly-established price for savings is allowed to remain remarkably indistinct.

The practical effectiveness of a particular monetary action is enjoyably debated. No one can tell what the effect of a given tightening of the money supply—with accompanying increase in the interest rate—will have. Often there can be considerable such tightening with no visible effect, and then there comes a sharp shrinkage in borrowing, investment and demand with, on occasion, disconcerting effect on output and employment. There is similar uncertainty as to the borrower's response to lower interest rates and easier borrowing terms. Unless this is accompanied by seemingly good prospects for selling goods or houses, nothing much may happen. Partly because of this uncertainty central bank actions always reflect the collective judgment of men who, individually, are in ignorance of the immediate consequences of their action. This and the intense, priestly and deeply lugubrious discussion which surrounds central bank policy serve, on the whole rather effectively, to disguise the uncertainty as to its effects.

Nonetheless it is basic to the neoclassical (and now neo-Keynesian) model that a combination of fiscal and monetary policy will produce comparatively stable prices at something close to full employment of the labor force. If there is unemployment, this can be offset by public action to increase demand. As unemployment disappears, inflation becomes the equal and opposite danger. This can be prevented by arresting the expansion in demand. With decent skill the price increases can be arrested with employment at a satisfactory level.

This optimism, it may be noted—and indeed stressed—is consistent with the neoclassical view of the market. The management of the economy works through the market. The producing firm is subordinate to the market. It follows that if total market demand is expanded, firms will respond sensitively to this instruction and increase output and employment. And —the vital point—if demand is contracted, they will respond by foregoing price increases or by reducing prices. The neo-Keynesian and the neoclassical faith are one; both depend on the same view of the power of the market.

3

The need to provide overall management of the economy greatly increased the role of the state in the economic system. This was not much emphasized in economic discussion. Nor was it supposed that this in any way altered the relation of the state to economic interest. The latter was still subordinate to the market; the overall management of the economy continued to be in response to citizen will and public purpose.

The nature of the development was favorable to this view. The original Keynesian action in the nineteen-thirties was readily and not inaccurately regarded as a humane response to the problem of mass unemployment. And the original design envisaged an increase in public expenditures for needed civilian activities of the government and therewith the increase in public expenditures over revenues that would add to aggregate demand. Social Security payments would also act automatically to increase demand. Once unemployment had been sufficiently reduced, it was imagined that expenditures could be reduced—or, as in the case of unemployment compensation, would reduce themselves. The activities of the government would then return to their previous scale.

The evolution of Keynesian policy was along very different lines. Instead of public expenditures that were increased or decreased as the economy required, public spending was set at a high basic level. This, in turn, was supported by personal and corporate income taxes which would rise more than pro-

portionately as income rose and would fall more than proportionately as income fell. Thus they acted automatically to
restrict or enhance private income and expenditure—they had
an automatic stabilizing effect on demand. And additionally,
as required, tax rates were decreased or, with more difficulty,
increased. The large expenditures meant that the role of the
government in the economy was no longer small but very
large. And in practice many of the expenditures either served
the needs of private firms or procured their products. Military
and other technical products were the most prominent examples and came to command an impressive share of the federal
budget. This was not deemed to be the result of the influence
of the firms so affected. To the extent that the state bought
arms it was in response to the general public interest as perceived by the citizen and reflected and interpreted by the legislature to the executive. "Peace reigns supreme in the realm of
neoclassical economics."[1]

The insulation of the overall management of the economy
from questions as to the influence of the firm—from the
thought that it could be an extensive adaptation to the needs
of modern corporate enterprise—kept faith with the simple
origins of the ideas. Supporting this tendency was the division
of labor in economics. This, as noted, is an old problem. The
tendencies of the modern corporation and union have never
been brought fully to bear on the theory of the firm, for they
are another branch of teaching and research. And the neoclassical view of the firm and the market is separated from
concern for the overall management of the economy by a yet
wider chasm. The theory of the firm is microeconomics; concern for the overall management of the economy is macroeconomics. Each part commands its own courses, instructors
and books. It is a distinction without meaning if macroeconomic policy reflects the needs of the modern corporation—
and this, indeed, we shall see to be the case. But the distinction

[1] Harry Magdoff, "Militarism and Imperialism," *The American Economic
Review,* Papers and Proceedings, Vol. 60, No. 2 (May 1970), p. 237.

exists and while doing so serves to direct attention from the influence of the corporation on the larger policy.

So it remains that in the neoclassical model the individual—more precisely, as we shall see presently, the household—remains supreme both in the private economy and the state. And so accordingly does the social, moral and political sanction thus accorded to the society. Two further matters remain to be mentioned.

4

In their response to consumer and public demand firms in the neoclassical model are, save for one feature, homogeneous—the model has only one theory of the firm. Large or small, firms respond in their development to market and consumer instruction. None has a special tendency to command capital and go ahead on its own; being wholly subject to the instruction of the market, none has the power to do so. The exception lies, as noted, with oligopoly or monopoly. But here investment and growth are still determined by what will maximize profits, and that is determined by the demand for the monopolist's or oligopolist's products. The only difference—one astonishingly in conflict with all present anxieties, as we shall see—is that where firms are powerful in their markets, there will be a smaller investment, a lesser use of labor, a lower level of development than would be socially ideal.

There remains the possibility that some firms or some industries may make greater use of technology—have a higher rate of technical innovation—and for this reason have a higher rate of development than others. On this the neoclassical model is ambiguous. It would be agreed that some industries are technically more progressive than others, but as to the reason there is no settled explanation. One line of argument, descending from the late Joseph A. Schumpeter,[2] holds that

[2] See Joseph A. Schumpeter, *Capitalism, Socialism and Democracy* (New York and London: Harper and Brothers, 1942), pp. 81 ff.

oligopoly and monopoly are technically more progressive than competitive enterprise. Because of their monopoly profits they can spend more for technical development; they are encouraged to do so because their monopoly power allows them to keep for themselves more of the resulting gains. The contrary and more conventional view is that firms with monopoly power are likely to be backward; they use their power to repress or suppress invention. An old economic cliché holds that the monopolist yearns for nothing so much as a quiet life.

Perhaps the most general view is that technical progress is random. It occurs as someone perceives a consumer need that has not been filled or sees a better way of making a product or rendering a service that fills a present need. (Thus technical innovation, like all else, is ultimately in response to consumer will.) If some parts of the economy are technically more progressive than other parts, it is because competition forces a greater mental activity in some areas as compared with others.

5

It will be guessed that the neoclassical system is not a description of reality. And this the ensuing pages will affirm. On what does its hold on the economic mind depend?

That it performs an instrumental service in guiding attention away from inconvenient fact and action has already been stressed. Accordingly it is a formula for a quiet noncontroversial life. But this is not all—for economists, as for others, truth and self-respect have their claims. The neoclassical system owes much to tradition—it is not implausible as a description of a society that once existed. Nor is it entirely unsatisfactory as a picture of that part of the economy hereinafter called the market system.

Additionally it is the available doctrine. Students arrive; something must be taught; the neoclassical model exists. It has yet another strength. It lends itself to endless theoretical refinement. With increasing complexity goes an impression of increasing precision and accuracy. And with resolved perplexity goes an impression of understanding. If the economist is sufficiently "caught up in his data and his techniques," he can overlook social consequences—his attention being elsewhere, he can even, without damage to conscience, "support a system that maltreats large numbers of people."[3]

It should not be supposed, however, that the present hold of the established or neoclassical system is secure. The link between doctrine and reality cannot be stretched too far. That the comparative development in housing and space travel is a manifestation of consumer will cannot be believed. Nor does anyone suppose that there is a tendency to equality in wage income as between different sectors of the economy. When belief is stretched too far, it snaps; the doctrine is rejected. The same is true of refinement without relevance. It comes, sooner or later, to seem but a game.[4] Not surprisingly in recent years the neoclassical model has been losing its hold—especially on the minds of younger scholars.

One consequence of the rejection of the neoclassical model is a renewed interest in Marx. The Marxian system was once the great alternative to classical economic thought. Numerous of its tenets are in striking contrast with the more implausible assumptions of the neoclassical model. It accords a major role to the large enterprise. That enterprise and its owner, the capitalist, do not lack power. Their superior technical competence is also granted. So is their tendency to combine into fewer units of ever greater size—the tendency to capitalist con-

[3] John G. Gurley, "The State of Political Economics," *The American Economic Review*, Papers and Proceedings, Vol. 61, No. 2 (May 1971), p. 53.
[4] ". . . the achievements of economic theory in the last two decades are both impressive and in many ways beautiful. But it cannot be denied that there is something scandalous in the spectacle of so many people refining the analysis of economic states which they give no reason to suppose will ever, or have ever, come about . . . It is an unsatisfactory and slightly dishonest state of affairs." F. H. Hahn, former president of the Econometric Society, cited by Wassily Leontief, in his Presidential Address to the American Economic Association, 1970, "Theoretical Assumptions and Nonobserved Facts," *The American Economic Review*, Vol. 61, No. 1 (March 1971), p. 2.

centration. The capitalists are not subordinate to the state; the state is their executive committee.

This reaction is not one, as the ensuing pages make clear, in which I concur. Marx saw much of the tendency of capitalist development, but he did not have the supernatural power of seeing in his time all that would eventually transpire. Much has happened since Marx of which account must now be taken. But because he was so long forbidden to honest thought, honesty and courage are now associated with the full acceptance of his system. This is to substitute one insufficient view of economic society for another. Honesty and perhaps also courage are associated with acceptance of what exists.

Consumption and the
Concept of the Household

> They were so *busy*—busy shopping, chauffeuring, using their
> dishwashers and dryers and electric mixers, busy gardening,
> waxing, polishing, helping with the children's homework, col-
> lecting for mental health and doing thousands of little chores.
>
> —Betty Friedan
> *The Feminine Mystique*

IN THE NEOCLASSICAL SYSTEM consumption is a generally flaw-
less thing to be maximized by any honest and socially benign
means. It is also a curiously trouble-free enjoyment. Thought
must be given to the selection of goods and services. No problems arise in their use. None of this is true, and what is omitted
from view deeply shapes the patterns of individual, family and
social life. This omission, and the circumstances which lie
back of this myopia, must now be examined. They are matters
of no small consequence.

Beyond a certain point the possession and consumption of
goods becomes burdensome unless the tasks associated therewith can be delegated. Thus the consumption of increasingly
elaborate or exotic food is only rewarding if there is someone
to prepare it. Otherwise, for all but the eccentric, the time so
required soon outweighs whatever pleasure is derived from
eating it. Increasingly spacious and elaborate housing requires
increasingly burdensome maintenance and administration. So
also with dress, vehicles, the lawn, sporting facilities and other
consumer artifacts. If there are people to whom responsibility

for administration can be delegated and who, in turn, can recruit and direct the requisite servant labor force, consumption has no limits. Otherwise the limits on consumption are severe. In looking at the great houses of seventeenth-, eighteenth- and nineteenth-century England, the first thought is of the wealth of the inhabitants. Often it was modest by modern standards. More should be attributed to the ability to delegate administrative responsibility for consumption to a large, willing and disciplined servant class.

Personal service has always been threatened by the more attractive labor opportunities provided by industrial development. It is also made more necessary by the wealth that such development provides. Not surprisingly, therefore, much effort has been devoted in the past hundred years to finding ways of preserving it or in finding surrogates for it or in devising substitutes. The search for surrogates has led generally to women and the family. It has made use of a pervasive force in the shaping of social attitudes—one that has often been sensed but rarely described. A name for it is needed, and it may be called the Convenient Social Virtue.

2

The convenient social virtue ascribes merit to any pattern of behavior, however uncomfortable or unnatural for the individual involved, that serves the comfort or well-being of, or is otherwise advantageous for, the more powerful members of the community. The moral commendation of the community for convenient and therefore virtuous behavior then serves as a substitute for pecuniary compensation. Inconvenient behavior becomes deviant behavior and is subject to the righteous disapproval or sanction of the community.

The convenient social virtue is widely important for inducing people to perform unpleasant services. In the past it has attached strongly to the cheerful, dutiful draftee who, by accepting military service at rates of pay well below the market, appreciably eased the burden of taxes on the relatively well-to-do taxpayer. Anyone resistant to such service was condemned as deeply unpatriotic or otherwise despicable. The convenient social virtue has also helped obtain the charitable and compassionate services of nurses, custodial personnel and other hospital staff. Here too the resulting merit in the eyes of the community served as a partial substitute for compensation. (Such merit was never deemed a wholly satisfactory substitute for remuneration in the case of physicians.) Numerous other tasks for the public good—those commonly characterized as charitable works—are also greatly reduced in cost by the convenient social virtue. But the convenient social virtue has been most useful of all in solving the problem of menial personal service.

In the last century and earlier in the present one the household domestic was regularly pictured as a person uniquely worthy of esteem. Nothing reflected more admirably on a person than diligent and enduring service to another. The phrase "old family retainer" suggested merit only slightly below that of "wise and loving parent." The phrase "good and faithful servant" had recognizable scriptural benediction. In England a large and comparatively deft literature associated humor, conversational aptitude, social perception and great caste pride with a servant class. None of this, however, stemmed the erosion to industrial employments. The ultimate success of the convenient social virtue has been in converting women to menial personal service.

In preindustrial societies women were accorded virtue, their procreative capacities apart, for their efficiency in agricultural labor or cottage manufacture or, in the higher strata of the society, for their intellectual, decorative, sexual or other entertainment value. Industrialization eliminated the need for women in such cottage employments as spinning, weaving or the manufacture of apparel; in combination with technological advance it greatly reduced their utility in agriculture. Meanwhile rising standards of popular consumption, combined with the disappearance of the menial personal servant, created an urgent need for labor to administer and otherwise manage consumption. In consequence a new social virtue came to attach to household management—to intelligent shopping for goods, their preparation, use and maintenance and the care and maintenance of the dwelling and other possessions. The virtuous woman became the good housekeeper or, more comprehensively, the good homemaker. Social life became, in large measure, a display of virtuosity in the performance of these functions—a kind of fair for exhibiting comparative womanly virtue. So it continues to be. These tendencies were already well advanced in the upper-income family by the beginning of the present century. Thorstein Veblen noted that "according to the ideal scheme of the pecuniary culture, the lady of the house is the chief menial of the household."[1]

With higher income the volume and diversity of consumption increase and therewith the number and complexity of the tasks of household management. The distribution of time between the various tasks associated with the household, children's education and entertainment, clothing, social life and other forms of consumption becomes an increasingly complex and demanding affair.[2] In consequence, and paradoxically, the menial role of the woman becomes more arduous the higher the family income, save for the small fraction who still have paid servants. The wife of the somewhat senior automobile executive need not be intellectually alert or entertaining, although she is required to be conventionally decorative on occasions of public ceremony. But she must cook and serve her husband's meals when he is at home; direct household procurement and maintenance; provide family transport; and, if required, act as charwoman, janitor and gardener. Compe-

[1] Thorstein Veblen, *The Theory of the Leisure Class* (Boston: Houghton Mifflin, 1973), p. 128.
[2] For a brilliant analysis of the time requirements of consumption (and also much else) see Staffen Burenstam Linder, *The Harried Leisure Class* (New York: Columbia University Press, 1970).

tence here is not remarked; it is assumed. If she discharges these duties well, she is accepted as a good homemaker, a good helpmate, a good manager, a good wife—in short, a virtuous woman. Convention forbids external roles unassociated with display of homely virtues that are in conflict with good household management. She may serve on a local library board or on a committee to consider delinquency among the young. She may not, without reproach, have full-time employment or a demanding avocation. To do so is to have it said that she is neglecting her home and family, i.e., her *real* work. She ceases to be a woman of acknowledged virtue.

3

The conversion of women into a crypto-servant class was an economic accomplishment of the first importance. Menially employed servants were available only to a minority of the preindustrial population; the servant-wife is available, democratically, to almost the entire present male population. Were the workers so employed subject to pecuniary compensation, they would be by far the largest single category in the labor force. The value of the services of housewives has been calculated, somewhat impressionistically, at roughly one fourth of total Gross National Product. The average housewife has been estimated (at 1970 wage rates for equivalent employments) to do about $257 worth of work a week or some $13,364 a year.[3] If it were not for this service, all forms of household consumption would be limited by the time required to manage such consumption—to select, transport, prepare, repair, maintain, clean, service, store, protect and otherwise perform the tasks that are associated with the consumption of goods. The servant role of women is critical for the expansion of consumption in the modern economy. That it is so generally

[3] Ann Crittenden Scott, "The Value of Housework: For Love or Money," *Ms.* magazine, (July 1972).

approved, some recent modern dissent excepted, is a formidable tribute to the power of the convenient social virtue.

As just noted, the labor of women to facilitate consumption is not valued in national income or product. This is of some importance for its disguise; what is not counted is often not noticed. For this reason, and aided by the conventional pedagogy as presently observed, it becomes possible for women to study economics without becoming aware of their precise role in the economy. This, in turn, facilitates their acceptance of their role. Were their economic function more explicitly delineated in the current pedagogy, it might invite inconvenient rejection.

4

The neoclassical model has, however, a much more sophisticated disguise for the role of women. That is the household. That the model emphasizes the role of individual decision in the economic system has been sufficiently stressed. This moral sanction would be seriously eroded if that decision depended on the facilitating toil of women—and if the decision-making role of women were seen to be subordinate to that of men.

These difficulties are elided by the concept of the household. Though a household includes several individuals—husband, wife, offspring, sometimes relatives or parents—with differing needs, tastes and preferences, all neoclassical theory holds it to be the same as an individual. Individual and household choices are, for all practical purposes, interchangeable.[4]

[4] Although some scholars have been troubled. "In the theory of demand we take the household as our basic atom . . . we should notice that many interesting problems concerning conflict within the family and parental control over the fate of minors are ignored when we take the household as a basic decision-making unit. When economists speak of *the* consumer, they are in fact referring to the group of individuals composing the household." Richard G. Lipsey and Peter O. Steiner, *Economics,* 2d ed. (New York: Harper and Row, 1969), pp. 71–72. Having taken note of the heroic simplification in identifying the individual and the household, the authors revert to tradition and let the simplification stand. They are, however, exceptional in mentioning the problem.

The household having been made identical with the individual, it then distributes its income to various uses so that satisfactions are roughly equal at the margin. This, as observed, is the optimal state of enjoyment, the neoclassical consumer equilibrium. An obvious problem arises as to whose satisfactions are equated at the margin—those of the husband, the wife, the children with some allowance for age or the resident relatives, if any. But on this all accepted theory is silent. Between husband and wife there is evidently a compromise which accords with the more idyllic conception of the sound marriage. Each partner subordinates economic preference for the greater pleasures of propinquity and the marriage bed. Or only individuals with identical preference schedules marry. Or in a hitherto unnoticed instrument of the marriage sacrament these schedules are made equal thereafter. Or, if the preference schedules do differ, divorce ensues, and the process continues until persons with identical schedules are mated. Or the woman, who in practice does much of the buying, equalizes her preferences at the margin, and her husband contrives to live in a lesser state of satisfaction. Or the husband, as the dominant member of the family, makes the decisions in accordance with his preference schedule, and his wife, however resignedly, goes along.

In fact the modern household does not allow expression of individual personality and preference. It requires extensive subordination of preference by one member or another. The notion that economic society requires something approaching half of its adult members to accept subordinate status is not easily defended. And it is not easily reconciled with a system of social thought which not only esteems the individual but acclaims his or her power. So neoclassical economics resolves the problem by burying the subordination of the individual within the household, the inner relationships of which it ignores. Then it recreates the household as the individual consumer. There the matter remains. The economist does not invade the privacy of the household.

5

The common reality is that the modern household involves a simple but highly important division of labor. With the receipt of the income, in the usual case, goes the *basic* authority over its use. This usually lies with the male. Some of this authority is taken for granted. The place where the family lives depends overwhelmingly on the convenience or necessity of the member who makes the income. And both the level and nature or style of expenditure are also extensively influenced by its source—by whether the recipient is a business executive, lawyer, artist, accountant, civil servant, artisan, assembly-line worker or professor. More important, in a society which sets store by pecuniary achievement, a natural authority resides with the person who earns the money. This entitles him to be called the *head* of the family.

The administration of the consumption resides with the woman. This involves much choice as to purchases—decisions as between different cake mixes and detergents. The conventional wisdom celebrates this power; it is women who hold the purse strings. In fact this is normally the power to implement decisions, not to make them. Action, within the larger strategic framework, is established by the man who provides the money. The household, in the established economics, is essentially a disguise for the exercise of male authority.

This household could not be better designed to facilitate consumption. The broad decisions on the general style of life rest with the husband and can be made without personal concern for the problems of administration that are involved. These are the business of his wife. Most things, including consumption, are more enjoyed if the work associated therewith is performed by someone else.

Women acquiesce normally in the crypto-servant functions of consumption administration—in arranging maintenance and repair of the house and of the household machinery and of the automobile and other equipment, in procurement and preparation of food, in supervision of the consumption of the young, in organization and management of social enjoyments, in participation in competitive social display. Such tasks are taken to be the natural responsibilities of the sex. It will be urged that there is no cause here for comment or complaint; most women perform these functions contentedly and even happily.

In a more comprehensive view the acceptance and happiness are an impressive tribute to the social conditioning to which people are subject. It is a prime tenet of modern economic belief—one that is central to the established economics and powerfully reinforced by advertising and salesmanship—that happiness is a function of the supply of goods and services consumed. This point having been established, how better can a woman contribute to her own happiness and that of the family she loves than by devoting herself to the efficient and energetic administration of the family consumption? Her service to the economy thus hitchhikes on her sense of duty and her capacity for affection. And, as with other economic needs, it is affirmed in the convenient social virtue. This celebrates as uniquely moral the woman who *devotes* herself to the well-being of her family; is a gracious helpmate; is a good manager; or who, at lesser levels of elegance, is a good housekeeper or real homebody. By comparison mere beauty, intellectual or artistic achievement or sexual competence is in far lower repute. And qualities that are inconsistent with good and acquiescent household administration—personal aggressiveness, preoccupation with personal interests to the neglect of husband and family and, worst of all, indifferent housekeeping—are strongly deplored.

In few other matters has the economic system been so successful in establishing values and molding resulting behavior to its needs as in the shaping of a womanly attitude and behavior. And, to summarize, the economic importance of the resulting achievement is great. Without women to administer it, the possibility of increasing consumption would be sharply circumscribed. With women assuming the tasks of administration, consumption can be more or less indefinitely increased. In very high income households this administration becomes, as noted, an onerous task. But even here expansion is still possible; at these income levels women tend to be better educated and better administrators. And the greater availability of divorce allows of a measure of trial and error to obtain the best. Thus it is women in their crypto-servant role of administrators who make an indefinitely increasing consumption possible. As matters now stand (and for as long as they so stand), it is their supreme contribution to the modern economy.

The General Theory of
Advanced Development

IN THE NEOCLASSICAL MODEL oligopoly—the market shared by
a few firms—is the only concession to the existence of the
large firm. In fact it reflects a small step in a giant process
which moves much of economic life dramatically away from
this model. The oligopolist can fix prices, control production.
But much more than this is involved when firms become large;
in fact a transformation of the very nature of economic society occurs.

The critical instrument of transformation is not the state or
the individual but the modern corporation. It is the moving
force in the change. But all social life is a fabric of tightly
interwoven threads. The change of which the corporation is
the driving force is a complex process in which many things
are altered at the same time and in which cause becomes
consequence and cause again. No description is uniquely correct; much depends on where one breaks into this matrix.[1]
But a starting point which has application over the whole
development is technology and its yet more important counterpart which is organization.

Technology—the development and application of scientific
or systematic knowledge to practical tasks—is a central feature of modern economic development. It comes to bear on

[1] See *The New Industrial State*, 2d ed., rev. (Boston: Houghton Mifflin, 1971),
p. 45.

both products and services and on the processes by which these are made or rendered. Organization goes hand in hand with technical advance. Little use can be made of technology from the knowledge available to any one man; all but invariably its employment requires the shared knowledge of several or numerous specialists—in short, of an organization. But this is only the beginning. To make technology effective, capital is required—plant, machinery, assembly lines, power, instrumentation, computers, all the tangible embodiment of technology. The management of this equipment also requires specialists and more organization.

With rare exceptions the more technical the process or product, the greater the gestation period that is involved—the greater the elapse of time between the initial investment and the final emergence of a usable product. Goods being in process for a greater time, the investment in working capital is greater. Steps must be taken to ensure that initial decisions are not ruined and capital lost by events that occur before the results are achieved. The capital that is now at risk and the organization that now exists must be paid for—are an overhead cost. It is incurred or persists whatever the level of output. This adds to the need to control intervening events. Things that might go wrong and jeopardize sales and therewith the return to capital or the revenue that is needed to pay for organization must be prevented from going wrong; things that need to go right must be made to go right.

In specific terms this means that prices must, if possible, be under control; that decisive costs must also be under control or so managed that adverse movements can be offset by the controlled prices; that effort must be made to ensure that the consumer responds favorably to the product; that if the state is the customer, it will remain committed to the product or its development; that other needed state action is arranged and any adverse government action prevented; that other uncertainties external to the firm are minimized and other external needs assured. In other words the firm is required, with increasingly technical products and processes, increasing capital, a lengthened gestation period and an increasingly large and complex organization, to control or seek to control the social environment in which it functions—or any part which impinges upon it. It must plan not only its own operations; it must also, to the extent possible, plan the behavior of people and the state as these affect it. This is a matter not of ambition but of necessity.

For any given level and use of technology there is, no doubt, a technically optimum size of firm—the size which most economically sustains the requisite specialists, the counterpart organization and the associated capital investment. But the need to control environment—to exclude untoward events —encourages much greater size. The larger the firm, the larger it will be in its industry. The greater, accordingly, will be its influence in setting prices and costs. And the greater, in general, will be its influence on consumers, the community and the state—the greater, in short, will be its ability to influence, i.e., plan, its environment.

More important, as organization develops and becomes more elaborate, the greater will be its freedom from external interference. In the small, uncomplicated enterprise authority derives from the ownership of capital—of the means of production. In the large and highly organized firm authority passes to organization itself—to the technostructure of the corporation.[2] At the highest level of development—that exemplified by the General Motors Corporation, General Electric, Shell, Unilever, IBM—the power of the technostructure, so long as the firm is making money, is plenary. That of the owners of capital, i.e., the stockholders, is nil.

As organization acquires power, it uses that power, not surprisingly, to serve the ends of those involved. These ends —job security, pay, promotion, prestige, company plane and private washroom, the charm of collectively exercised power

[2] See Chapter IX.

—are all strongly served by the growth of the enterprise. So growth both enhances power over prices, costs, consumers, suppliers, the community and the state and also rewards in a very personal way those who bring it about. Not surprisingly the growth of the firm is a dominant tendency of advanced economic development.

This growth, with the associated exercise of power, is the primal force by which economic society is altered. In its practical manifestation, however, it is singularly uneven. In some parts of the economy such growth by the firm is subject to no clear upper limit. In other parts it is subject to severe limits or it proceeds against increasing resistance. Where the growth is stunted, so, of course, is the capacity to persuade consumers as to products, and the state as to products and needs, and so is the technical competence that goes with organization. These are facts of the first importance for understanding the modern economy. This is why, in some parts of the economy, production and the associated blessings are great or excessive and in other parts deficient. It is why the rewards to workers and other participants are far more favorable in some parts of the economy than others. It explains, as we shall see, much else.

2

As noted, the normal thrust for growth is aborted or stunted in numerous industries. This is a fact of prime importance, and the point at which growth is arrested is exceptionally clear. It is where direction by an individual, either an owner or his immediate agent, would have to give way to direction by an organization. Some tasks can be performed by organization, some cannot. In those industries where organization is inapplicable or ineffective the firm remains at a size which allows its operations to be performed or guided by a single individual. Four factors exclude organization, make necessary individual performance or direction.

Organization is excluded where the task is unstandardized and geographically dispersed. In such a case central supervision cannot easily or economically be maintained, and, the scale of operations at each geographical point being necessarily small, no very sophisticated technology with associated capital equipment can be brought to bear. In these instances there is no substitute for the incentive which accords a primary share of the return (or the lack of it) to the skill, intelligence and effort expended by an individual. Adding to the advantage of the individual in such instances is the frequent opportunity for exploiting himself and on occasion his family or immediate employees. Organizations are subject to rules as to pay and how hard or how long people are worked; individuals in relation to themselves or their families are not. For this reason they can flourish where organizations do not.

The second factor confining the firm to the authority of an individual person is the surviving demand for explicitly personal service. Where one person pays for the personal attention of another person, technology is ordinarily limited or excluded. Organization has few or no advantages.

The third factor limiting the scale of the firm is involvement with art. Scientists and engineers lend themselves well to organization. Although professional vanity celebrates their inspired individual creativity, they function normally in teams and with considerable and costly equipment which also requires management. The artist lends himself much less well to organization. Accordingly, if the product or service involves original and genuine (as distinct from repetitive or banal) artistic expression, the firm will always be small. Frequently, as in the case of personal services, it will be identical with an individual.

Finally, the firm is on occasion kept small by law, professional ethos or trade union restriction which prohibits the technology or organization (e.g., group medical practice) that would allow of the growth of the firm. This has especial application to the professions and the building trades, although in both cases it exists in combination with the geographical dispersion of the activity which also limits the size of the firm.

The chapters following return to the operation of the foregoing restraints on the growth of a firm.

3

The combination of a powerful thrust to the expansion of the firm in some parts of the economy with effective restraints on growth in other parts produces a remarkably skewed pattern of economic development. This is manifest in all nonsocialist industrial countries; the skewness is also evident in the Eastern European states and the Soviet Union. In the United States one may think of one thousand manufacturing, merchandising, transportation, power and financial corporations producing approximately half of all the goods and services not provided by the state. In manufacturing the concentration is greater. The two largest industrial corporations, General Motors and Exxon, have combined revenues far exceeding those of California and New York. With Ford and General Electric they have revenues exceeding those of all farm, forest and fishing enterprises. In the first quarter of 1971, the 111 industrial corporations with assets of a billion dollars or more had more than half of all the assets employed in manufacturing and received substantially more than half of the earnings on more than half of the sales. The 333 industrial corporations with assets of more than $500 million had a full 70 percent of all assets employed in manufacturing.[3] In trans-

[3] Testimony of Willard F. Mueller, Hearing before the Select Committee on Small Business, United States Senate, 92d Congress, 1st Session, November 12, 1971, p. 1097. The inclusion of unconsolidated assets would increase the share of these corporations in all industrial assets. Summarizing, Professor Mueller observes in the same testimony, "there exists an extremely asymmetrical industrial structure, with the bulk of economic [i.e., industrial] activity controlled by an elite of a few hundred enormous corporations and the remainder divided among four hundred thousand small and medium-sized [manufacturing] businesses."

portation, communications, power utilities, banking and finance, although the concentration is somewhat less, the tendency is similar; in merchandising the concentration is also high. An assembly of the heads of the firms doing half of all the business in the United States would, except in appearance, be unimpressive in a university auditorium and nearly invisible in the stadium.

Making up the remainder of the economy are around twelve million smaller firms, including about three million farmers whose total sales are less than those of the four largest industrial corporations; just under three million garages, service stations, repair firms, laundries, laundromats, restaurants and other service establishments; two million small retail establishments; around nine hundred thousand construction firms; several hundred thousand small manufacturers;[4] and an unspecified number serving the multivariate interests of an advanced society in what is collectively called vice.

No agreed level of assets or sales divides the millions of small firms which are half the private economy from the handful of giant corporations which are the other half. But there is a sharp conceptual difference between the enterprise that is fully under the command of an individual and owes its success to this circumstance and the firm which, without entirely excluding the influence of individuals, could not exist without organization. This distinction, which may be thought of as separating the twelve million small firms from the one thousand giants, underlies the broad division of the economy here employed. It distinguishes what is henceforth called the market system from what is called the planning system.

4

That the planning system does not conform to the neoclassical model—that its firms are not passive in response to the

---

[4] *Statistical Abstract of the United States, 1972,* U.S. Department of Commerce. Figures are for 1969.

market or the state—will not be difficult to establish. Mainly it is a matter of breaking with accustomed and stereotyped thought. To this part of the economy we will return. The market system, in its admixture of monopoly and competition, does conform in broad outline to the neoclassical model. The latter model is a rough description of half of the economy; it has lost touch with the other and, in many ways decisive, half. Precisely because of its capacity for radical change the nonmarket part has transformed itself into something very different. But the market system also departs from the neoclassical model in two respects: Intervention by the state in this part of the economy is more extensive and altogether more normal than the theory suggests. And the market system must exist alongside the planning system. As might be imagined, its development is powerfully affected by this latter fact.

Subject to the constraints of knowledge, energy and ambition the firm in the market system, competitive or monopolistic, does maximize its profits. For this there is an affirmative incentive; in contrast with the firm in the planning system, where organization has taken power from the owner, the man in charge gets the profit or, at a minimum, can be tested and rewarded in accordance with his ability to produce profits. But the negative motivation may be even more compelling. If profits are good, the tendency of the firm will be to expand. Others will do likewise. In the usual case yet others can enter the business, for the required capital, in keeping with the small size of the firm, is also small. And, in contrast with the firms in the planning system, those already in the field do not have the advantage of a ready-built organization. All of this is to say that a little monopoly is far harder to protect than a big one. So, in the market system, production and prices are not likely to be effectively and reliably under the control of the firm. Nor are they likely to be subject to the collective authority of a few firms. So, if profits are abnormal, they will soon come down. This means that the entrepreneur does not for long have the luxury of preoccupying himself with any goal except that of making money. He must always, where this is concerned, do the best he can. Amateur defenders of the market, enchanted to discover, as did Adam Smith two centuries ago, that good seems to proceed from evil, have often gone on to conclude that avarice is an original virtue. This is to make virtue of what, in fact, is necessity.

It follows also from the absence of control over prices and production that in the market system much of the egalitarian tendency of the neoclassical system survives.

Since in the market system earnings are not likely long to be above the level that compensates the entrepreneur for his effort and capital, there will be no very reliable source of savings from business earnings. Accordingly the firm will be dependent—as the firm in the planning system is not—on outside sources of capital. This is a circumstance of much importance, as we shall see when we come to consider the public regulation of the economy. Should this involve the regulation of borrowing, as it does, the market system will be affected with particular force.

The firm in the market system can, by itself, do little to influence the behavior of its customers. The resources are lacking for the effort. Additionally the farmer who tried as an individual to recruit customers for his wheat, cattle or tomatoes would be charitably shaping consumer preferences for all the producers of these staples since wheat, cattle and tomatoes are largely indistinguishable whatever their source. And all know this to be so, as they may not in the case of gasoline. This homogeneity of product, coupled with the unappealing scale of their operations and income, is why no farmers are seen on Madison Avenue.

As the individual member of the market system cannot typically influence his customers,[5] so he cannot influence the state. The president of General Motors has a prescriptive

[5] Agriculture is the purest case of the firm that is powerless in this regard. In the service industries, as I note presently, the firm does have some relation to its customers.

right, on visiting Washington, to see the President of the United States. The president of General Electric has a right to see the Secretary of Defense and the president of General Dynamics to see any general. The individual farmer has no similar access to the Secretary of Agriculture; the individual retailer has no entrée to the Secretary of Commerce. It would be of little value to them if they did. The public bureaucracy, as we shall later see, can be effectively and durably influenced only by another organization. And between public and private organizations there can be a deeply symbiotic relationship.

Innovation in the market system also conforms generally to that depicted by the neoclassical model. That means that it is very limited. Most innovation requires that there be capital to cover the period of development and gestation and to pay for the equipment which is its counterpart.[6] This the firm in the market system lacks; even more significantly it lacks the specialized technical and scientific talent and counterpart organization which modern technical development all but invariably requires. No important technical development of recent times—atomic energy and its applications, modern air transport, modern electronic development, computer development, major agricultural innovation—is the product of the individual inventor in the market system. Individuals still have ideas. But—with rare exceptions—only organizations can bring ideas into use. Innovation in the market system remains important mostly in the minds of those who cannot believe the small entrepreneur ever fails.

5

While the firm in the market system is subject to the market constraints and disciplines of the neoclassical model, it does

[6] See Edwin Mansfield, "Innovation and Size of Firm," in *Monopoly Power and Economic Performance*, Edwin Mansfield, ed. (New York: Norton, 1964), pp. 57–64.

not accept them with any pleasure. We may lay it down as a firm rule that all participants in the economic system will wish to modify such constraints in their own favor. They will try to influence prices, costs, consumer decisions and the actions of the community and the state. And this will be as true of the market system as the planning system. The difference is not in intent but ability. The market and its disciplines are greatly praised by scholars. They are rarely applauded by those who are subject to them.

Some limited independence of market constraints is inherent in the geographical dispersion of economic activity, the small volume of activity at any particular location and the high serviceability of the incentive system associated with individual entrepreneurship. This dispersion means, very often, that there is *room* in a particular area for only one or a few entrepreneurs. More neighborhood drugstores, pizza merchants, laundromats, and all would starve. Thus the firm acquires a measure of control over prices and production. And through the charm of his personality or the modest eloquence of his persuasion the proprietor may develop a certain hold on his customers. Instead of competition there is differentiation of product or service by its association with the personality of a particular seller.[7] It is, needless to say, a highly circumscribed control—a motorized and mobile population is uniquely able to escape any effort at exploitation by its neighborhood monopolist.

The neoclassical model takes product differentiation and local monopoly more or less in stride. It is less tolerant of collective efforts at market control. These are numerous and frequently invoke the assistance or initiative of the state. The worker rejects the opportunity of selling his services on the market as an individual and unites with others to sell them

[7] See Edward H. Chamberlin's *The Theory of Monopolistic Competition* (Cambridge: Harvard University Press, 1933). Market control which depends on such product differentiation was called by Chamberlin "monopolistic competition."

through the union. The union thus gains power over the common price of such services and, through control of apprenticeship or membership, on occasion over supply. Government support to collective bargaining reinforces this control. The small clothing manufacturer or builder uses the union scale which is common to all plus a conventional markup as the basis for pricing his product. So do others, and all thus win control (sometimes tenuous) over prices. Physicians, chiropractors, osteopaths, optometrists, lawyers and the building trades control or influence supply by control of educational requirements, apprenticeship requirements or state licensing. Farmers persuade the state to fix prices by government purchase and, through acreage or marketing quotas, to limit supply. Small manufacturers seek publicly enforced retail price maintenance, small retailers protection against preferential treatment of large competitors under the Robinson-Patman Act. All of these efforts reflect the tendency of all producing firms, whether in the market or planning system, to control their economic environment and not be subject to it.

In agriculture this effort has gone beyond the control of production and prices to embrace somewhat elementary efforts to influence consumer response. The nutritional and moral benefit from consuming milk and its products is advertised. Similarly fruits, nuts and other agricultural specialties. In recent times the effort of the United States Department of Agriculture in promoting the consumption of tobacco has been in interesting contrast with the effort of the Department of Health, Education and Welfare to portray its lethal effects.

In agriculture there has also been a notably successful effort to exempt technology from the constraints imposed by the market system. This has been accomplished (as also we shall see in the planning system) by the socialization of technical innovation and its propagation—this being the accomplishment of the federal and state experiment stations and laboratories, the colleges of agriculture and the agricultural extension services. The planning system has also contributed heavily to technical innovation in agriculture through the farm equipment and chemical industries. So, in lesser degree, have corporations which are directly participant in agriculture either through contracts with farmers for feeding poultry or livestock or by direct operations, as in the case of fruit and vegetable growing. Those who cite modern agriculture as an example of the progressive tendencies of small enterprise and the market economy invariably overlook the role of the government and the supplying corporations. No innovation of importance originates with the individual farmer. Were it not for the government and the farm equipment and chemical firms, agriculture would be technologically stagnant.[8]

6

The difference between the planning and the market systems does not lie in the desire to escape from the constraints of the market and to effect control over the economic environment. It is in the instruments by which these are accomplished and the success with which they are attended. Participants in the market system who want stabilization of their prices or control of their supply must act collectively or get the assistance of the government. Such effort is highly visible and often ineffective, unsuccessful or inefficient. Voluntary collective efforts can be destroyed by a few deserters. Legislators do not

[8] As noted earlier in this chapter, the large firm is impelled to control its prices (and other aspects of its environment) in order to protect the investment that technology requires. This is also one of the important services to agriculture of government price-fixing. Such price stabilization makes it possible for farmers to invest in the working capital and equipment that such technology requires in a manner that would never be possible were they subject to the vagaries of unmanaged prices. This does much to explain the great increase in agricultural productivity since farm price support legislation was initiated in 1933. This intervention does not accord with the instruction of the neoclassical model—it puts prices above the equilibrium level and keeps them from clearing markets. In consequence it is deplored as unsound policy and as a source of social inefficiency. This criticism is regularly voiced by scholars who praise the efficiency and technical progressiveness of the American farmer.

always respond—even to farmers. If action is taken, it is often taken apologetically, for it is recognized that the established economics disapproves.

In the planning system, in contrast, the firm wins its control over prices automatically and with no public fanfare merely by being large. Similarly over output. And it can be large because its tasks lend themselves to organization. There are also things for which it needs the support of the state. But its approach is not to the legislature but to the bureaucracy. This is more reticent. And, the bureaucracy being more powerful, it is also likely to be more effective. The consequence, not surprisingly, is that firms in the market system get much attention for accomplishing very little by way of setting aside market constraints or otherwise modifying the economic environment to which they are subject. And the large firms in the planning system get no attention for accomplishing much. This is extensively reflected in economic pedagogy. The political power and depredations of the farm lobby are much celebrated by economists. The far more powerful control by General Motors of prices, costs and consumer responses and its far more influential association with the Department of Transportation, the Department of Defense and the regulatory agencies are much less remarked.

7

The planning system seeks to exercise control over its economic environment, and, as later chapters sufficiently establish, it succeeds. The market system manifests the same desire, is much more visible in its effort and is much less successful. The one system dominates its environment; the other remains generally subordinate to it.

But the planning system is very much a part of the environment to which the market system is subordinate. It supplies power, fuel, machinery, equipment, materials, transportation, communications that the market system uses. It also supplies a large share of the consumers' goods and services that participants in the market system purchase. And it is an important purchaser of the products of the market system—most notably in the case of agriculture. A prime feature of this relationship will already be evident. The market system buys at prices which are extensively subject to the power of the planning system. And an important part of its products and services are sold at prices which it does not control but which may be subject to the market power of the planning system. Given this distribution of power, there is a prima facie case that things will work better for the planning system than for the market system. The terms of trade between the two systems will have an insouciant tendency to favor the system that controls its prices and costs and therewith the prices and costs of the other system as well. A further effect, unless there is unimpeded mobility between the two systems, will be inequality of return—a relatively secure and favorable income for participants in the planning system, a less secure and less favorable return for those in the market system. To these hypotheses I will return, for, alas, they have solid substance. First it is necessary to examine more carefully the central characteristics of the two systems.

TWO

THE MARKET SYSTEM

VI

Services and the Market System

SERVICES ARE RIGHTLY ASSUMED to be the domain of the small firm and thus of the market system. In recent times there has been much talk in the United States and other industrial countries of the rise of the so-called service economy. This, in turn, has been taken by determined defenders of the market to prove that the market-controlled economy is not only surviving but resurgent. Economics, as it is taught, is being saved from the depredations of the great corporation by the growing demand for services.

On examination this development turns out to be a good deal more complex. Numerous service enterprises are the by-product of the rise of the large firm. They are, in effect, a subsidiary and supporting development of the planning system. This is especially the case with that part of the service sector which, from outward evidence, is expanding most rapidly.

Services, nonetheless, are the favored domain of the small firm. As mentioned in the last chapter, the growth of the firm is arrested where the task is geographically dispersed and where, accordingly, the activity at any one point is limited and where the task involved is unstandardized. This means that one or a few people work in isolation, i.e., without supervision. Under such circumstances they adopt their preferred pace, which is normally slow. They enhance their input of mental and physical energy only if, in their earnings, they reap the rewards and suffer the penalties of the individual entrepreneur.

Geographical dispersion, it may be observed, is not an absolute barrier to organization. If the task is relatively standardized, performance norms can be established for dispersed workers to which they can then be required to conform. Or payment can be according to product or revenue produced. Or such dispersed functions can be associated with the capital and technical support of a larger organization, as in the case of a local unit of a retail or restaurant chain. In recent times there has also been a great exfoliation of hybrid arrangements, commonly denoted franchises, by which an individual is made responsible for a local enterprise and is thus subject to the comprehensive incentive system that is associated with individual entrepreneurship. Ordinarily he is required to put some of his own capital at risk. As the proprietor he is then rewarded for any and all energy and intelligence he expends and punished for failure in this regard, as well as for such errors of optimism and gullibility as he may have made and such other misfortune as he may endure. At the same time the parent corporation provides the advertising, capital and technical assistance (both real and imagined) that the individual could not, as a wholly independent businessman, provide for himself. Nonetheless geographical dispersion of unstandardized tasks remains a general bar to the growth of the firm.

2

Geographically dispersed activities fall into two classes: those that by their nature require area, such as agriculture, or those that involve personal service. If it is a matter of importance and can be afforded, the customer will travel some distance to obtain a service, as in the case of divorce, abortion or a visit to the Mayo Clinic. But, in the main, services must be located in relation to those who use them. This is in contrast with manufacturing, which, more often, is located in proximity to materials, an experienced or available labor force, the economies that are associated with the presence of like operations as in the case of dress manufacture or the accident of where the industry got started, as with rubber in Akron or cars in Detroit.[1]

In addition, while many services can be rendered impersonally by organizations and increasingly this is the tendency, others are specifically valued for their association with the personality of the purveyor. No technology is involved; an increase in the amount of the service requires a proportional increase in the time of those rendering it. No or slight advantage accrues from organization. Such is the case, more or less, with the ministrations of physicians, psychiatrists, lawyers and prostitutes. In all these industries the small enterprise survives.

3

However the most rapidly proliferating service enterprises are found, paradoxically, where machinery is replacing personal service, including the service of the personal servant. This substitution brings into existence numerous new dispersed and unstandardized tasks which, accordingly, lend themselves well to the small firm. This development is part of a much larger change.

In the preindustrial past a very large part of nonagricultural economic activity consisted in *personal* service by one individual to another. The preparation and serving of food; the administration of the wardrobe; assistance in personal adornment and hygiene; provision of education, entertainment and religious solace; physical protection of the person; accommodation of sexual need and numerous other services by one individual immediately to another were of such sort. The individual rendering the service, priestly officers and mistresses sometimes excepted, stood in a menial relationship to the user

[1] Natural rubber was never produced in the Akron vicinity nor are its products uniquely consumed thereabouts. The case of cars and Detroit is much the same.

of the service. A practiced servility was, itself, an attribute of the service. Livery and even the praise accorded the well-trained and thus suitably self-abnegate servant affirmed the inferiority.

Industry, from its earliest appearance, had a deeply inimical impact on menial personal service. However oppressive and grim in its earlier stages the factory provided an environment in which the individual was not required to advertise or acknowledge an inferior status in relation to another person. He shared an impersonal servitude with many others. In time, also, since factory work lent itself to mechanization and collective action to obtain a larger share of the resulting increase in productivity, the pay of factory workers came to exceed that for personal service. The factory worker had not only more dignity but more money as well.

Paralleling and in considerable measure consequent upon the escape of the domestic worker to the factory were two further developments. One, already examined, was the conversion of women to a crypto-servant role in the household, a role made urgent by the increased volume of consumption now requiring administration. The other, to be noticed in a moment, was the transfer of numerous services hitherto performed in the household to the domain of the small firm and the independent entrepreneur.

Accompanying and also aiding the persuasion of women to their crypto-servant role has been the elaboration of a wide range of mechanical devices designed to aid them in the care and use of goods and the administration of consumption generally. Consumption having created the need for this labor of women, labor-saving devices followed. Washing machines, refrigerators, vacuum cleaners, automatic heating, a great variety of cooking equipment are all designed (and constantly redesigned) to minimize or offset the labor that is associated with increased consumption of goods.

These devices, in turn, sustain an increasing number of service enterprises. Many of the devices require professional installation. More require expert service and repair. These operations, being geographically dispersed, lend themselves to the incentive system of the small enterprise and, accordingly, are extensively performed by independent entrepreneurs. Additionally, as will be noted later, the planning system that supplies this equipment is oriented in its technology not to what is durable and not always to what is useful. Rather it is to what can be sold—to what lends itself to consumer persuasion. This persuasion trades extensively on the assumption that anything that is technically new reflects progress. The result is much innovation that emphasizes newness or novelty at the expense of durability or the competence of mechanical or engineering design. This increases the demand for the services of small, geographically dispersed firms that are associated with the repair and maintenance of these devices or the installation of new devices which, being of novel design, the buyer can be persuaded are better.

A high level of consumption, in combination with the burdens of the crypto-servant role, has also led to the contracting out of consumption from the household to the independent entrepreneur. This continues an ancient trend. The physician, priest, schoolmaster, concubine and prostitute originally had household status. All, like the servant turned factory worker, early escaped to the more civilized status of independent worker. In more recent times the burdens of high consumption have had a similar effect. The washing of clothes by the housewife has passed extensively to the laundry and laundromat; the cooking of food has passed similarly to the restaurant; food that is still consumed at home is precooked or otherwise pre-prepared; dishes, table linen and numerous other household artifacts are provided by service establishments, perhaps to be discarded after use; houses are cleaned and gardens maintained by external contractors. Such services, being also unstandardized and geographically dispersed, lend themselves to the incentive system of the small firm.

4

In the conventional economic view a large range of consumer products are defended for the way they economize domestic toil—for their effect in lifting the burden of household labor from the housewife. This and the prospect for her greater ease and leisure have, in turn, been counted one of the civilizing effects of modern industrial development. To express skepticism on the association between goods and happiness is to be reminded, sometimes sternly, that one does not realize what the dishwasher "has meant" to the average woman.

It will now be evident that the conventional view is both disingenuous and reflects a considerable oversimplification. That higher standards of consumption have increased the demand for personal services cannot be doubted. The most urgent requirement has been for a full-time household servant and manager. And, since industrial development eliminates the menial servant class, this role, under the pressure of the convenient social virtue, has been assumed by the crypto-servant wife. Household equipment and externally rendered services then act to mitigate the labor associated with this role. If the crypto-servant function of women is taken as the point of departure, there can then be no question that mechanical household gadgetry, as also services external to the household, greatly ease this role. But clearly one should look more deeply at the sources of the toil that is so mitigated. One must begin with the high-consumption economy and the need for a crypto-servant class for its administration.

For the moment, however, we may let these matters stand. It is sufficient that with economic development, and consequent social change, the service sector of the economy does survive and expand—and that this is the result, in very important part, of the development of the planning system and the need to administer, facilitate and service its consumption. In further consequence opportunity for the small entrepreneur and firm will continue. And so, pari passu, will this part of the market system.

VII

The Market System and the Arts

AS SERVICES RESIST organization, so also do the arts. This has not been much noticed, and the oversight is not remarkable. Economics has never had a serious view of art. Science and technology are important matters. Painting, sculpture, music, the theater, design are more frivolous. The manufacture of canvas, paint or pigment is a worthy concern of the economist; anything that lowers the cost of these commodities or expands their output contributes to economic goals. But the quality of the painting as distinct from the paint or what causes artists to colonize, multiply and prosper has never been thought a proper concern of the subject. Artistic achievement may, in principle, be part of the claim of an age or place to development. But as compared with the production of goods or technical or scientific accomplishment it has no practical standing. None of this is an accident. The relevant attitudes are firmly grounded in the nature of modern economic society.

2

The artist is, by nature, an independent entrepreneur. He embraces an entire task of creation; unlike the engineer or the production-model scientist he does not contribute specialized knowledge of some part of a task to the work of a team. Because he is sufficient to himself, he does not submit readily to the goals of organization; to do so is to sacrifice his view of what is artistically worthy to the organization view. That is to sacrifice artistic integrity, for the latter, whether the result be good or bad, is always coordinate with the artist's own view of his task.

Not needing the support of organization and not being able or allowed to accept the goals of organization, the artist fits badly into organization. As so often, everyday speech reflects the fact. The excessively independent man is regularly defended in an organization as "something of an artist." The brilliant nuisance or unemployable eccentric is said to be "altogether too much of an artist." The artist, on his side, finds life in any sizable and successful organization cloying, restrictive or even stifling. And he is required, if he is to have the good opinion of his peers, to say so.

In consequence, except in some slight measure where the discipline of organization is itself artistic—that of a symphony orchestra or a ballet company—the artist functions as an independent entrepreneur (a term he is unlikely to employ) or, as in the case of a self-respecting architect, as a member of a very small firm which he can dominate or in which he can preserve the identity of his work. A few industries—the motion picture firms, television networks, the large advertising agencies —must, by their nature, associate artists with rather complex organization. All have a well-reported record of dissonance and conflict between the artists and the rest of the organization. A small literature—Budd Schulberg's *What Makes Sammy Run?*, Evelyn Waugh's short story, "Excursion in Reality," Rod Serling's *Patterns*—celebrates this conflict and the crudities of organization men as viewed by artists. Frequently the problem is solved by removing actors, actresses, script writers, directors, composers, copywriters and creators of advertising commercials from the technostructure of the film studio, television network or advertising agency and reconstituting them in small independent companies. The large firm then confines itself to providing the appropriate facilities for producing and—more importantly—marketing, exhibiting or airing the product. Simi-

larly painters, sculptors, concert pianists and novelists[1] function, in effect, as one-man firms or, as in the case of rock,
dance and folk music groups, as small partnerships and turn
to larger organizations to market themselves or their product.

3

Where manufacturing requires a measure of artistic effort and
is judged in part by this, the artistic superiority of the small
firm will often allow of its survival in competition with the large
organization. Since the good artist cannot or will not be subordinate to organization, the large, relatively immobile enterprise
commands not the best talent but the most accommodating.
This, more or less by definition, is second-rate. Nor is this
purely a matter of poor or perverse taste on the part of the
organization. The large firm must have designs that will lend
themselves to long and economical production runs. Artistic
sense must also yield to the will of those who, on the basis
of instinct, experience or market research, are knowledgeable
on what the public can be persuaded to buy. Artistic judgment
is subject to a supervening view of acceptability, and this, in
turn, is powerfully influenced by the common working assumption, sometimes articulated, that no one ever went into receivership underestimating the popular taste. In consequence the
large firm gets long runs, technical efficiency, low costs and
a considered marketing strategy at the expense of good design.
The automobile industry, the mass producers of furniture, the
household appliance industry, the container industry and numerous others amply illustrate the point.

In the smaller manufacturing firm in which the artist has
a dominant role or in which, at a minimum, the discipline of
organization is less severe, there is more scope for individual
assertion which is essential. In consequence design can be much

---

[1] When the late Ian Fleming, the manufacturer of James Bond, turned himself
into a limited company a short while before his death, it was a matter of
worldwide comment.

better. Further, the artist having a dominant role, the design will not be subordinate to efficiency on the production line. It will reflect the artist's sense of what is good, not the technician's knowledge of what can be efficiently produced or the marketing man's sense of what can be sold. Thus, although technically less proficient, the small firm has, because of its small size, the advantage of superior art. In the manufacture of apparel, jewelry, timepieces, furniture, other household artifacts and in cooking, house construction and publishing this advantage can be considerable. Invariably the small firm serves what is called the upper end of the market, i.e., that providing a more expensive product to more affluent consumers who have superior taste or (perhaps the more common case) superior guidance thereto.

On occasion small firms are kept in being by large firms which need, but cannot themselves employ, the talent the small firms command. The big dress manufacturers buy models from the small designers; the automobile manufacturers seek the help of Italian entrepreneurs; DuPont turns to small firms in Paris and New York for fabric designs. It is easy to hire chemists, a DuPont official observed in conversation a few years ago, and to know what you are getting. But no one knows how to hire good artists, and they won't live in Wilmington, Delaware.

The small firm also derives an advantage from the distinctive nature of the demand curve for the work of the artist. The position of this demand curve—the amount that people will buy at any given price—is a function of time. People, as presently to be emphasized, are persuaded strongly to believe that technical innovation is a *good* thing—that it is coordinate with progress. This being so, the market response to such innovation is generally favorable. Such attitudes, of course, coincide with and reflect the needs of the planning system. Public attitudes toward artistic innovation, by contrast, have been subject to no such conditioning. Accordingly the first impression of artistic innovation is almost invariably unfavorable. What is new is commonly thought offensive or grotesque. So it was with the Impressionists, the Cubists, the Abstract Expressionists, and so it is with the modern exponents of Pop Art. The situation is similar in prose, poetry and much music. It follows that the initial market for innovative work in the arts is almost always small; only as taste develops does demand expand. But some are attracted and some have pleasure in seeming to appreciate what others reject. Thus they are willing to pay. This situation—a small market in which cost is secondary to the quality of the artistic achievement—also lends itself well to the individual or the small firm.[2]

4

In the past one of the most common manifestations of affluence was expenditure on the arts. The quality of the public and ecclesiastical architecture and its embellishment and of the public entertainment was the visible test of community achievement. For the private household the excellence of the dwelling and its painting, sculpture, furniture, food and conversation provided a similar test. This was especially true of those communities—Venice, Florence, Genoa, Amsterdam, Bruges and Antwerp—where the orientation was strongly to economic achievement. Military and sexual prowess, accomplishment in courtly routines and manners and commitment to gastronomic and alcoholic excess have always been the principal rivals to the arts as manifestations of civilized accomplishment. The commercial cities, as compared with the courts, were generally more resistant to all these forms of display.

In modern times the arts, as a measure of both community and private achievement, have undergone a great relative decline. Scientific and engineering accomplishments have become

[2] While the contrast between the popular reaction to technical and to artistic innovation depends on social conditioning, this may not be the whole explanation. It may be that visual reactions are inherently conservative and then, with time, undergo accommodation. For this reason new grotesqueries in dress or automobiles—those for which no one would dream of advancing any artistic claim—become after a time visually tolerable.

overwhelmingly more important, and these have also invaded the esteem that was once associated with military prowess. Few now speak of the discipline, parade-ground virtuosity or bravery of soldiers, sailors or aviators. It is the excellence of their tanks, nuclear submarines, aircraft or the guidance systems with which they are equipped that excites comment and measures national accomplishment. Space exploration is an even more dramatic example of the use of scientific and engineering virtuosity as a measure of national achievement. As medieval towns once compared the magnificence of their cathedrals and the munificence of their decoration, modern superpowers parade the number, conception and cost of their expeditions, manned or unmanned, to the moon and the planets or their laboratories in orbit around the earth. The reward, however, continues to be partly metaphysical and spiritual.[3] It is part of the normal argument for scientific or engineering expenditure that there is great ultimate human benefit. In the case of the exploration of the moon it is generally agreed that there was little or none. That we did not ask for such benefit was, in this instance, a measure of our intellectual and spiritual maturity. Again we see the influence of the convenient social virtue.

Scientific and engineering accomplishments are likewise the accepted measure of achievement in other fields—physics, chemistry, genetics, aviation, computer technology. No one would think of according similar importance to the comparative accomplishments of the Soviet Union and the United States in painting, the theater, the novel or industrial design. At least

[3] In his prepared remarks for the return of the first astronauts from the moon Dr. George C. Mueller, space-flight director for NASA, urged Americans not to "substitute temporary material welfare for spiritual welfare and long-term accomplishment." He went on to plead that we "dedicate ourselves to the unfinished work so nobly begun by three of us to resolve that this nation, under God, will join with all men in the pursuit of the destiny of mankind . . ." The power of such spiritual commitment should not however be exaggerated. Dr. Mueller subsequently accepted a higher salary as a vice president of General Dynamics. See Richard F. Kaufman, *The War Profiteers* (Indianapolis and New York: Bobbs-Merrill, 1970), p. 80.

until recently, in any competitive display of painting, poetry or music, both countries would have been compelled to delete their best or most interesting offering. Americans, selecting art for such display, would have had to avoid what numerous congressional critics would condemn as Communist-inspired work. The Soviets would have had similarly to exclude work that reflected bourgeois decadence. Since engineering and scientific accomplishments are the accepted measures of community achievement, it follows that education and other support for these subjects is not only a proper but a highly desirable employment of public funds. The arts, for obvious reasons, have no similar claim.

No one will be in doubt as to the source of these attitudes. It lies with the technostructure and the planning system and with their ability to impose their values on the society and the state. The technostructure embraces and uses the engineer and the scientist; it cannot embrace the artist. Engineering and science serve its purpose; art, at best, is something which it needs but finds troublesome and puzzling. From these attitudes come those of the community and the government. Engineering and science are socially necessary; art is a luxury.

5

While artistic accomplishment has ceased to be a test of social achievement, the esoteric and pretentious apart, it has a continuing and perhaps increasing importance for the individual and the household. The everyday standards for assessing the reputability and general social position of a family do eschew any artistic element. They center, instead, on the supply of standard goods. The inhabitants of a three-bedroom house are thought "better off" than those of a two-bedroom house; they gain further distinction from having a fully equipped kitchen and being a two- as opposed to a one-car family. The technical character and novelty of goods, not their beauty, is stressed in advertising. To attack the design of a consumer product as banal is often to invite an indignant response. It is what the people want. The critic is an elitist.

However at higher income levels an artistic sense or pretense in domestic architecture, interior decoration, furniture, landscape design and even in food and entertainment tends either to be enjoyed for its own sake or is part of the claim to esteem. This, in turn, sustains a substantial and growing market for the work of artists as well as for individuals who provide guidance to those who, often perceptively, lack confidence in their own taste. As a consequence an appreciable volume of modern economic activity depends not on the technical efficacy of the product or the efficiency with which it is produced but on the quality of the artists associated with the design. Some industries are so based. Danish and Finnish furniture owe their modern distinction not to engineering competence but to artistic worth. Italian products excel not in engineering but in appearance. The postwar renaissance of Italian industry had a similar basis. And there is a similar if less visible development in the United States. It is as yet little recognized—no one would think of encouraging the artist as opposed to the engineer, scientist or business manager as the foundation of future industrial development. But its monopoly of artistic achievement provides an important assurance of the survival of the small firm.

6

Over a longer period of time the arts and products that reflect artistic accomplishment will, for the foregoing reasons, be increasingly central to economic development. There is no reason, a priori, to suppose that scientific and engineering achievements serve the ultimate frontiers in human enjoyment. At some point, as consumption expands, a transcending interest in beauty may be expected. This transition will vastly alter both the character and structure of the economic system.

It will first have to contend, however, with the social conditioning of the technostructure and the planning system which, as just noted, relegate to a minor social role what cannot be embraced or used. The transition will have also to contend with the convenient social virtue of the artist. This requires a word, for it is what causes the artist to accept an inferior economic and social role both for himself and for art in general.

Specifically the artist has been persuaded that the world of economics is one to which, by nature, he has little relation. It is part of his pride that the number who can appreciate the work of the true artist—who have a valid response to its meaning—must always be very small. So his market and the resulting compensation must be meager. This meagerness, in turn, shows his merit. The more deprived his life, the more truly is he an artist. Only the more modest religious offices share the artist's belief that merit is inversely related to compensation.

This view of the artist of himself has two social advantages. It economizes expenditure on the arts, for, if pecuniary reward does not improve and possibly even damages performance, it should obviously be kept to the minimum. And it means that all but a minority of artists are kept safely in that subordinate and anonymous state that is reserved for the indigent or near-indigent. They do not in consequence compete with managers, scientists or engineers for social esteem. Nor do they compete with scientists and engineers for public funds to support the arts.

The claim on public resources is further reduced by the belief, also accepted in varying measure by the artist, that not much can be done about education in the arts. Given only the money, any number of scientists and engineers can be supplied. These can be processed from almost any lumber. The number of artists produced cannot, however, exceed the number of people with intrinsic talent, and the part of the population with such talent is assumed, although on no known evidence, to be small. And it is part of the folk attitude toward the arts that the truly inspired artist will excel, whatever the barriers to be overcome. The convenient social virtue thus minimizes the need for expenditure on education in the arts.

It is worth recalling that until about a hundred and fifty years ago the convenient social virtue held the scientist to be an unworldly and monkish figure whose support was properly the function of a private patron. The public dignity of the artist, being older and better established, was then far greater, and the artist had a strong claim on public resources. The scientist has long since escaped from these monkish origins; personal affluence and public support are no longer imagined to be damaging to his creative instinct. On the contrary they are deemed necessary for it. The artist, by contrast, still depends extensively on private patronage; he has come, along with the rest of the community, to accept the view that public support to the arts could be dangerously repressive of the artistic spirit.[4] It is obvious that the resulting saving in public funds—as compared with a society that sets the same store by the arts as, say, by moon travel—is very great.

7

Thus the arts. They will continue to be a major stronghold of the individual and the small firm. They will also be an expanding part of economic life. The opportunities for enjoyment from artistic development have no visible limit; they are almost certainly greater than those from technical development.

But this expansion would be much greater were the sources of our present attitudes on art, science and technology better understood. The arts now have an infinitely smaller claim than science and engineering on both private and public resources. This, we have seen, is the result not of public preference but of conditioned belief. People—including artists themselves—are persuaded to accord importance and priority to what is within

[4] Architects, being needed by industry, are emancipated from the belief that artistic achievement is damaged by economic association or personal affluence.

the competence and serves the needs of the technostructure and the planning system.

The means for emancipating belief—for releasing it from service to the planning system—is a matter to which, obviously, we must return.

VIII

Self-Exploitation and Exploitation

IN ORGANIZATION people work in accordance with prescribed rules. Hours of arrival and departure are specified; a minimum expenditure of effort is enforced throughout the working day. This last is accomplished by supervision, or by the establishment of norms which the worker must meet, or by the use of machinery (the notable case being the assembly line) to set the pace for those who serve it, or by the use of piece rates and incentive systems to grade pay to specific quantitative performance.

A central function of the modern trade union is to bring the rules to which the worker is subject partly within his jurisdiction—to ensure that, however indirectly, he has a say in their nature and enforcement. This means that, for many in the organization, not only minimum but also maximum expenditure of effort is specified—is subject to regulation. Depending on the point of view this latter regulation is greatly praised for its humanizing effect on modern industry or greatly condemned as an arbitrary limitation on worker productivity.

It may be noted that rules specifying or limiting effort diminish steadily in importance as one goes upward in an organization hierarchy. For white collar workers performance is extensively ensured by making diligent and eager effort identical with decent and approved behavior. An organization where such behavior is general is spoken of as having high morale. Anything that contributes to such morale—cheerful, acquiescent cooperative effort—has a high standing in the convenient social virtue. In the higher reaches of organization rules disappear and are replaced by the struggle for competitive self-advancement or, perhaps more important, the desire for the fear, respect or approval of one's peers. These, in turn, are earned, at least in part, by visible contribution to the purposes of the organization. The common consequence is that as the individual becomes exempt from rules requiring effort, he expends not less but more such effort. He must be unsparing of himself in his hours. He impresses himself and others by the intensity of his thought and effort during his working day. His leisure, so-called, is spent or said to be spent in contemplating his responsibilities or in purely therapeutic or business-supporting recreation. In eccentric cases an executive may claim never to take a problem home with him. But this is rare. A conspicuous emphasis on effort is almost always considered to be a sound career strategy. A man needs to be known as a hard-driving executive.[1]

In the small firm rules, as a way of ensuring effort, become unimportant. They give way to the incentive system of the individual entrepreneur which rewards him comprehensively for effort and punishes him comprehensively for sloth and ineptitude. And, in the case of his few employees, instead of formal regulations there is personal supervision. Such a way of ensuring effort is especially useful, as earlier noted, for dispersed and unstandardized tasks for which performance rules cannot easily be formulated. And it is also extremely useful where, as in the case of many services, the success may depend more on the subjective reaction of the customer than on the energy or technical skill committed to the task. Thus the ability of the operator of the gasoline station, motel or lunch counter

---

[1] In the university, by contrast, the convenient social virtue emphasizes a much more reserved approach to toil. The ruminative and even mildly unreliable man is valued. It is an aid to scholarly reputation to take long, restorative vacations. The excessively busy professor risks being thought insufficiently reflective, of using busyness as a cover for some scholarly deficiency or, at minimum, of being unaware of the need to conserve scarce mental energy. A consummately lazy professor often greets a normally diligent colleague with the admonition, "Aren't you overdoing it? You must be more careful."

to hold his natural surliness below the level of overt insult— or even to convey an aspect of affable obeisance commonly called "being obliging"—may be more important than effort or technical competence. This is best ensured by making him personally subject to the rewards and penalties of his behavior.

To dispense with rules governing the minimum level of effort is also, obviously, to dispense with rules limiting the maximum expenditure of effort. This means that, some flexible inhibitions of law and custom apart, nothing regulates the hours of work of the individual entrepreneur, and nothing at all regulates the intensity of his effort. He may thus be in a position to offset the higher technical productivity of the better-equipped worker in the organized but regulated sector of the economy by working longer, harder or more intelligently than his organization counterpart. In doing so, he reduces his compensation per unit of effective and useful effort expended. He is, to put the matter differently, almost wholly free, as the organization is not, to exploit his labor force since his labor force consists of himself. The term exploitation, it should be noted, is here used in its precise sense to describe a situation in which the individual is induced, by his relative lack of economic power, to work for less return than the economy generally pays for such effort.

Self-exploitation is extremely important for the survival of the small firm. It is vital in agriculture. It is of the utmost importance for the small or one-man enterprise in other fields— retailing, restaurants, repair services, household services and the like. So useful is it as a way of inducing high-quality service at low cost that a number of the Communist countries—Poland and Hungary as well as Yugoslavia—allow private entrepreneurship, with associated opportunity for self-exploitation, in gasoline retailing, restaurants, cafés and other service trades. In these countries such services are more conveniently, expeditiously and economically performed than in the Soviet Union, where such self-exploitation is discouraged. Not surprisingly self-exploitation by the small entrepreneur has a high standing in the convenient social virtue.

2

The notion of exploitation in all popular thought is associated with the employee; exploitation by the employer, or self-employed entrepreneur, of himself has been accorded much less recognition. It seems possible that it is of more economic and social importance than the similar treatment of employed labor. In practice, however, self-exploitation and the exploitation of employed labor go hand-in-hand in the modern economy.

As noted, the small employer ensures effort by employees not by establishing rules but by personal supervision. And, as no rules keep this employer from reducing his own reward for given effort, he strongly resists any regulation that keeps him from similarly depressing that of his employees. He feels a natural license to ask of others what he asks of himself.

These tendencies are most clearly visible in agriculture. Self-exploitation by the farmer of himself and his family has long been greatly approved behavior—a powerful manifestation of the convenient social virtue to which I will return. Along with much mention of harvest, weather and the peculiarities of crop production this has become the basis for the farmer's claim to the right similarly to exploit his labor force. In the United States this claim has been all but universally conceded; the farmer is normally exempt from legislation on wages and hours, and trade unions in agriculture are specifically denied the support of the National Labor Relations Act. (This exemption from legal regulation and immunity from unions also extend, of course, to very large farmers where the counterpart self-exploitation is not evident.)

Along with the farmer the small-town tradesman, the small manufacturer or processor or other small employer has also been a point of strong resistance to unions, wages and hours legislation, social security legislation and other regulation of conditions of work. Large firms, far more closely associated with exploitation in the imagery of social thought, have been far less resistant. This has been a puzzle to all who dwell on the surface of matters. Why should the good small man be so bad? Usually it has been concluded that a minimal grasp of social issues goes naturally with a minimal scale of operations— or that there is something peculiarly retarding about any association with the soil. We see that, as usual, the actual explanation is well grounded in economic circumstance. The small entrepreneur, being comparatively powerless in his market, cannot with certainty pass higher wage costs or benefits along to the public in his price. And he correctly senses that he survives by being able to reduce the wage that he receives for the effort he expends. He seeks to retain the same right as regards those whom he employs. Thus his resistance to unions, minimum wage legislation or whatever might increase his wage costs.

3

The large corporation is the recipient of little social praise. The small entrepreneur, in contrast, is all but universally admired. Part of this is social nostalgia; the small businessman is the modern counterpart of the small firm of the classically competitive economy. As such, he is a reminder of a simpler and more comprehensible world. But more of the praise, without doubt, reflects the convenient social virtue. What is praised is what serves the comfort and convenience of the community.

Not all that is so praised withstands close examination. Thus the small entrepreneur is hailed as a man of rugged independence. That this independence is often circumscribed both in principle and in practice by a strenuous struggle for survival goes unmentioned. He is said, in contrast with the organization man, to be admirably unfettered in his political and social views. As just noted, these are likely, out of necessity, to be an uncompassionate reflection of self-interest. Living outside of organization, he is said to rejoice in freedom from the discipline of organization. No one gives him orders; no one supervises his work. He can look any man in the eye. It is not noted that this is often the caution, conformity, obeisance, even servility, of a man whose livelihood is at the mercy of his customers. His is often the freedom of a man who is pecked to death by ducks.

In the large corporation no one doubts that there should be limits on hours of toil, on the effort that must be expended and on all other conditions of work. The role of the unions in bringing and defending such humanizing rules is applauded. So is that of the state. But in the market system the small entrepreneur who rises early and works late, is available to his customers around the clock and is unremitting in the intensity of his toil, is the man who merits praise. No tedium marks his efforts; he is a public benefactor and a model for the young. Especially stalwart is the farmer who holds a job in town, works evenings, Sundays and holidays on his land and engages his wife and children likewise. Not only does he receive credit himself, but additional praise for his industry is assigned to the relevant Swedish, Danish, Norwegian, German, Scottish, Finnish or Japanese provenance. That such toil is compelled by the circumstances of the market system is not remarked. That it may damage the health of children and that in agriculture it involves the denial of unions, minimum wages, even workmen's compensation, to those who need their protection the most is unmentioned. Such is the authority of the convenient social virtue.

4

Thus the market system. In addition to the factors excluding organization that have been examined in these last three chapters, there are also some parts of the economy where overt restriction acts to keep the firm small. Lawyers, physicians (and until recently stock exchange houses) are required by law or professional ethic to operate as individual proprietorships or partnerships. In the past some states have outlawed the corporation in agriculture. Illegal or semilegal enterprises—

brothels, pornographers, drug peddlers, illicit gamblers—are, as a practical matter, denied the opportunities for growth allowed by the corporate charter. All of this tends to keep the firm in these areas of economic activity small, although, given the nature of the function or service, this would probably be so anyway.

For a half century or more there has been debate as to whether the small firm is fated to disappear—whether the ineluctable tendency of the economy is to enterprises of great scale. Defenders of the neoclassical orthodoxy have always been persuaded of the importance of the small firm to their system. It is the least ambiguous manifestation of the market economy. According to temperament the defenders have divided between those who have argued that the small firm is threatened and thus requires the energetic protection and support of the state and those who proclaim its future (and therewith their system) to be entirely secure.

We see that there are tasks—much agriculture, geographically dispersed services, those involving the arts—which do not lend themselves to organization. And even where organization might be possible, the entrepreneur, by reducing his own compensation, increasing his effort and—within limits—doing the same with his employees, can survive in competition with organization. So the small entrepreneur will continue. Nor is there any clear reason to expect that his share in total economic activity—the share that is not accomplished by organization—will decline. Later discussion will leave no doubt that the development in the market system will be inferior to that in the organized sector of the economy. But this could be in relation to a need for development in the market sector that is much in excess of that in the planning sector. In economics it is of no slight importance that one be able to think in relative terms.

That the market system survives, at least in part, because of its ability to reduce the reward of its participants leads on to an obvious and ominous conclusion. It is that there is a presumption of inequality as between different parts of the economic system. The convenient social virtue adds to this presumption by helping people to persuade themselves that they should accept a lesser return—that some part of their compensation lies in their social virtue. Needless to say, the presumption of inequality becomes much stronger if one part of the system has power over its prices and costs, and these, in turn, are the costs and prices of the other part. That means it can, in effect, force exploitation on the other part.

That there is such exploitation as between parts of the economy we shall see. Along with the unequal development just mentioned it is one of the primary reasons for examining the economy as not one system but two. But before pursuing such matters further it is necessary to look at the other half of the economy. The unsusceptibility of the task to organization bans growth for the myriad of firms. But the susceptibility of some tasks to organization allows of infinite growth to the few. To areas of such growth we now turn.

THREE

THE PLANNING SYSTEM

IX

The Nature of Collective Intelligence

In the advanced capitalist nations, new elites based on science
and technology are gradually displacing the older elites based
on wealth.

> —Robert L. Heilbroner
> *The New York Times Magazine*

ORGANIZATION is an arrangement for substituting the more specialized effort or knowledge of several or many individuals for that of one. For numerous economic tasks organization is both possible and necessary. The production of standardized products and services—automobiles, steel, power, telecommunications—allows of extensive geographical concentration of workers. There is no art or anything that need be so described. The product or service does not depend on its association with personality. Or if it does marginally, as in the case of air travel, that too lends itself to standardization as in the attire or hair of the stewardess or the rigorously prescribed statement that the presence of the passenger has been a source of "our pleasure." So organization is possible.

It is also necessary. The production of standardized goods and services requires specialists who have a technical knowledge of particular details of the processes and products involved or who can contribute knowledge as to possible alteration or improvement. The counterpart of specialization is always organization—organization is what brings specialists, who as individuals are technically incomplete and largely useless, into a working relationship with other specialists for a complete and useful result.

However technical specialists are only the most obvious examples of a species. With organization, increasing size becomes possible for the firm; with that goes increasing power in markets, over community attitudes and in relation to the state. For the exercise of this power—for product planning, to devise price and market strategies, for sales and advertising management, procurement planning, public relations and government relations—specialists are also needed. To perfect and guide the organization in which the specialists serve also requires specialists. Eventually not an individual but a complex of scientists, engineers and technicians; of sales, advertising and marketing men; of public relations experts, lobbyists, lawyers and men with a specialized knowledge of the Washington bureaucracy and its manipulation; and of coordinators, managers and executives becomes the guiding intelligence of the business firm. This is the technostructure. Not any single individual but the technostructure becomes the commanding power. "We believe that today's and tomorrow's management problems are so complex . . . that a team should always decide."[1]

Where the task does not lend itself to organization, the size of the firm is limited by what is within the energy and intellectual competence of a single person. This competence may be greater or less, but it is finite. When the task lends itself to organization, there is no set upper limit to the size of the firm. For reasons presently to be examined it can become very large. This part of the economy, accordingly, is characterized by comparatively few, very large firms. This is its most striking feature.

It is a feature, it must be stressed, which is remarkably unrevealed by neoclassical economics. In this model the firm seeks to maximize its profits. Its costs are given or largely given by circumstances external to itself. So also is the demand for its products. So also, at any given time, is the technology that is available to it. From these circumstances comes an optimal

[1] Hans Süssenguth, Executive Board of Lufthansa. Interview with Robert Spencer, *The American Way* (June 1972), p. 20.

scale of operations—one at which the difference between cost and price, when multiplied by sales, is greatest. A substantial part of economic pedagogy—most of the dimly remembered geometry and mathematics of the elementary text and course— tells how this ideal scale of operations is determined. For every enterprise there is such an optimum size—such a limit. It may be exceeded but only because the management is captured by a malignant and irrational tendency to giantism which causes it to seek size in conflict with interest.

For the behavior of the modern enterprise to be understood there must be a forthright rejection of this historic stereotype. It could only be true if the control of costs, prices, demand and technology were independent of the size of the firm. Should it be that as the firm becomes larger, it is better able to control its costs, its technology, its prices, the responses of its consumers or the government (were all these a dependent variable associated with size), the scale at which profits are maximized could obviously increase with the increasing size of the firm. To increase size and associated control over costs, technological processes, prices, demand and the state could become, then, one way of maximizing profits. And, as will presently be seen, profit maximization is not, in any case, the central goal of the technostructure. Above a certain profit threshold the members of the technostructure are better rewarded by growth itself. The accepted economics is a remarkable barrier to understanding the most basic tendency of modern economic society. That is for constituent firms to become vast and to keep on growing.

2

The firms in the planning system—since all are corporations, that term will henceforth be extensively used—are by no means homogeneous. At one extreme are relatively small corporations where organization is still elementary—those which are in the first stages of transition from the individually guided firm of the market system. At the other extreme are General Motors and, among utilities, the larger (in assets though not in sales) American Telephone and Telegraph. As one proceeds from the smaller corporation to the giants, the role of any single individual diminishes, the authority of organization increases. Among the very large corporations of some age—those I shall refer to as the mature corporations—the power of organization is plenary.

Not without a wrench can we imagine an organization, public or private, which isn't subject to the guiding hand of an identifiable individual. The intrinsic vanity of human personality, a yearning for simple and comprehensible arrangements with known saints and sinners and, as I shall have later occasion to observe, a coalition of vested intellectual interests all seek to associate power with specific persons. Two factors serve, with the increasing size of the firm and with the passage of time, to transfer power from the individual to the technostructure.

The first factor is the strongly authoritarian character of collegial decision. As noted, organization is the arrangement by which specialists combine their information to make decisions— decisions that require the knowledge, experience or intuition of several or many persons. This is the case with almost all important action. No one person has more than a fraction of the knowledge necessary for the design, production or marketing of a new automobile model, missile or detergent. It follows that no one person can decide whether one of these should be designed, developed, produced, or can be sold. It follows further that no one person can pass on the investment involved. These matters must be resolved by groups—typically by committees—in which each participant contributes the specialized knowledge of which he is the possessor.

It follows, further, that anyone who is not a party to this collegial decision-making but who, nonetheless, seeks to alter or interfere with its decisions will be doing so on the basis of inadequate knowledge. This will be true of an owner who shows up at infrequent intervals to review the decisions taken by the organization. Or a member of the board of directors. Or a superior in the organization who was not himself a participant in the decision-making group. It will, of course, also be true of someone from the union or the government. Such intervention being uninformed, as must be that of any individual, it will risk being damaging. Power, as a result, passes from the owners or their nominal representatives on the board of directors to the management of the corporation. And within the corporation, for the same reason, it passes down to the individuals who are actively participant in contributing their knowledge to the decisions.

In general both the number and the complexity of the decisions increase with the size of the corporation. So, accordingly, does the monopoly of the technostructure over the knowledge that decisions require. So, accordingly, does the power of the technostructure.

People who contribute their knowledge to group decisions will not be inclined to underestimate the quality of information they supply. They will be sensitive to the importance of the process—the meetings, memoranda, printouts, conversations— by which it has been assembled and tested and combined with the information of others. They will be sensitive, accordingly, to the uninformed intrusion of those above or outside, and they will resist intrusion. An individual can yield to the decision of another individual whom he knows to be more knowledgeable than himself. A group will sense that it cannot so yield. What is often called bureaucratic arrogance reflects, in fact, the need to exclude the even more arrogant individual who does not know what he does not know. The natural sympathy for the individual should not cause anyone to be misled as to what is involved. Group decision-making is authoritarian because, by sound instinct, it must protect itself from poorly informed outsiders, including those who are nominally in positions of power.

The second influence according power to the technostructure is inherent in the growth and aging of the corporation. The small corporation, in which a few stockholders subscribe capital and lodge authority in one person, is distinguishable only in legal detail, notably in the limitation of liability, from the firm that is owned and operated by an individual. With increasing size the stockholders become more numerous. The passage of time, abetted by inheritance, inheritance taxes, philanthropy, alimony and the tendency of nonfunctional owners or their trustees to diversify holdings, also disperses stockholdings. Accordingly the share of total stock, and the associated power, of any individual diminish. Stockholders, accepting the weakness of their position, become passive; they vote their proxies automatically for the management slate or they do not vote them at all. The directors come to realize that their power derives from the management and not from the stockholders. They confine themselves, accordingly, to a ritualistic approval of management decisions.

This change is progressive. That the control of the large corporations tends to pass from the owner to the manager first came to general notice in 1932 in the classic study by Adolf A. Berle and Gardiner C. Means, *The Modern Corporation and Private Property*. They then concluded that, of the 200 largest nonfinancial corporations in the United States, control of 88, or 44 percent, was by the management. Under ordinary circumstances no group of stockholders could muster enough votes to challenge the authority of the self-appointed heads of the firm. Thirty years later the same study, using the same criteria, was repeated by Robert J. Larner. He concluded that, of the largest 200 nonfinancial firms in 1963, not fewer than 169—84.5 percent—were by then securely in the control of their management.[2] "Almost everyone now agrees that in the large

[2] Robert J. Larner, "Ownership and Control in the 200 Largest Nonfinancial Corporations, 1929 and 1963," *The American Economic Review*, Vol. 56, No. 4, Pt. 1 (September 1966), p. 777 et seq. There are subjective elements in the definition of management control that have been extensively exploited in order to keep alive the power of the owner and capitalist. However the Berle and Means conclusions now enjoy general acceptance. And there is no reason to think that the Larner calculations, save that they derive less prestige from age and author, are any less reliable.

corporation, the owner is, in general, a passive recipient of income; that, typically, control is in the hands of management; and that management selects its own replacements."[3]

3

The power of management is not flaunted. Rather it is subject to elaborate disguise. A universally observed ritual requires obeisance to the nominal sources of authority. Elderly boards of directors, selected by the management and meeting infrequently to ratify actions on which they are not informed, are said to be a valuable source of wisdom and guidance. Thus their power. The natural respect accorded to age or incipient senility assists the illusion. This, the solemnity of the corporate liturgy, the modest compensation that is involved and the modest perception that is required will often persuade outside directors—those not members of the technostructure—that they have power. This illusion is heightened by the need to pass, routinely, on the appropriation or borrowing of money or otherwise on financial transactions or accounts; nothing better sustains an impression of omnipotence than association, however nominal, with large sums of money.[4]

Similarly within the technostructure. Board chairmen or presidents are presented with the careful decisions of subordinate groups in an atmosphere of such deference that those so honored often fail to see that their function is confined to ratification. All who serve in a bureaucracy, public or private, are instinctively accomplished in such ritual. In public bureaucracies there is, perhaps, a special skill. Presidents, Secretaries of Defense, Prime Ministers and Ministers are elaborately briefed

[3] Edward S. Mason, *The Corporation in Modern Society* (Cambridge: Harvard University Press, 1959), p. 4.
[4] "They meet once a month, gaze at the financial window dressing (never at the operating figures by which managers run the business), listen to the chief and his team talk superficially about the state of the operation, ask a couple of dutiful questions, make token suggestions (courteously recorded and subsequently ignored), and adjourn until next month." Robert Townsend, *Up the Organization* (New York: Alfred S. Knopf, 1970), p. 49.

on matters which they do not comprehend. This is but rarely to allow them to decide. Rather it is to give them the impression, and allow them to give others the impression, that the decision is theirs. So believing, they are somewhat less likely to feel the need to assert their authority which, being uninformed, would be dangerous or damaging.

Power is not diminished by its being attributed to someone else. Almost invariably it is enhanced and made easier of exercise. Nothing serves the technostructure better than to attribute unpopular or socially reprehensible action to higher authority. "Unfortunately it is the interest of the stockholder that we must consider." "I must answer to the board of directors." Thus it is possible to have the reality of power without the penalties.

4

Some qualifications are now in order—for the person who is resisting truth nothing is so convenient as the overstatement, which becomes a handle for assault on the whole proposition. Only in the largest corporations is the power of the technostructure plenary—only there has it worked itself fully to completion. And there, if the corporation is failing to make money, stockholders may be aroused—although individual stockholders will usually accept the cheaper option of selling out. Proxy battles in very large firms occur all but exclusively in those that are doing badly.

Also, as stockholders are dispersed by some forces, they are aggregated by others—notably by insurance companies, pension funds, mutual funds and banks. This does something to arrest the deterioration of the power of the stockholder. However the effect is easily exaggerated. The tradition of financial institutions in relation to management is usually passive. There is a sense of the danger of uninformed intervention.

In the smaller corporations of the planning system, especially if they are concerned with technically less complex processes or products, an individual—a superior officer in the managerial hierarchy or an important stockholder—can be informed. Thus he can have influence on decisions. And in both large corporations and small there are three ways in which such an individual can exercise power over decisions which he lacks the information to make himself.

First, he can change the participants in the decision-making —he can dismiss, move and replace those who are involved. Much of the authority of the modern manager is associated with such casting.

Second, he can propose a change in the conceptual framework in which decisions are made. He cannot safely decide to produce a new product or employ a new process or acquire a new subsidiary. This requires the informed participation of specialists who have or can obtain the requisite knowledge. But he can ask consideration of such a new product, process or venture. The ultimate decision will still depend on knowledge that is available only from a group. But individual imagination can suggest new areas for the application of such group knowledge. It is commonly held, and may well be true, that group decision-making has its special competence within familiar boundaries or parameters and is peculiarly incapable of breaking out of them.

Finally, an individual can, by appeal to professional help, appraise the competence of the organization that makes decisions and the quality of the decisions that are being made. He does not, as an individual, command the knowledge to do this. But he can get another organization to do so. This accounts, in modern times, for the great expansion of the management-consulting industry. Together with firms rendering specialized technical services such consulting firms had revenues in 1970 estimated at around a billion dollars. The largest of the consulting firms had themselves become corporations, and some had become units in conglomerates.[5]

[5] "Consultants Clash Over Ownership," *Business Week* (November 27, 1971), p. 66.

These powers—casting of personnel, directing the decision-making process into new areas and placing it under scrutiny—are the principal prerogatives of individuals in the modern corporation, of what is called leadership. In the mature corporation they are the sole prerogatives of the individual acting as an individual.

None of this exercise of authority, it will be observed, involves substantive decisions—what or how to produce or how to market a product or service. While the intervention here described can be uninformed and arbitrary and hence damaging, it is not inherently so. Casting or management reorganization can be a threat to the members of the technostructure, for it means that members may be fired, reassigned or retired. This, like uninformed intrusion on substantive decision, is something to be resisted. To protect against such intrusion, as we shall see presently, is also one of the prime goals of the technostructure.

5

That power in the mature corporation passes from the stockholder to the management has long been conceded, in practice as distinct from theory, by economists. And increasingly it is accepted that the goals of management may be different from those of the owners—that, as Dr. Robin Marris of the University of Cambridge, Professor William Baumol of Princeton and others have urged, there will be greater concern for the reliability of revenues and especially for growth of the firm.[6] The separation of ownership from control involves a sharp challenge to the assumption of profit maximization. The pursuit of profit in the neoclassical model is relentless and single-minded. Acquisitiveness, avarice and cupidity (not man's most saintly attributes) are, by a most fortuitous paradox, the source of energy which is then subordinate to the public command

[6] On this see *The New Industrial State,* 2d ed., rev. (Boston: Houghton Mifflin, 1971), especially Chapter XV.

and the public good. Yet when ownership is divorced from control, a grievous problem arises. The acquisitiveness, avarice and cupidity that so valuably motivate the system are supplied by the managers—the technostructure—and their fruits go to the owners. The managers are not beholden to the owners. Thus the system works because those who are most acquisitive are also most conscientiously determined to toil on behalf of others. Avarice is philanthropically in the service of others. This is the remarkable contradiction which the modern large corporation poses for neoclassical economics. It is a problem which neoclassical theory solves mostly by ignoring it. It resorts to the most useful of the intellectual conventions of the economist which is, when inconvenient facts are encountered, to assume them away. In all formal theory and most pedagogy the entrepreneur, uniting ownership and a beneficial stake in profits with active direction of the enterprise, is held to be the important case. The corporate reality is ignored. Discussion begins with the observation, "Profit maximization is, of course, assumed." As always, where profit maximization is concerned, there is seeming reassurance in the frank acceptance of human greed as a basic point of departure for economic analysis. No one can doubt the importance of greed.

Sometimes, however, as just noted, the separation of ownership from control is conceded to have an effect on corporate goals. There is, it is agreed, more emphasis on security of earnings and growth, less on profit maximization. The effect of this on neoclassical attitudes has been slight. The corporation prices for some combination of security, growth and profit. But what it can do is still circumscribed by the market—prices may be a little lower, sales a little larger than if profits were exclusively the goal, but no revolutionary significance is attached to the changes. And were the firm fully subordinate to the market, the effect of pursuing these alternative goals might not, in fact, be great.

If, however, with the rise of the great corporation goes the power extensively to enforce its will on the society—not only

to fix prices and costs but to influence consumers, and organize
the supply of materials and components, and mobilize its own
savings and capital, and develop a strategy for handling labor,
and influence the attitudes of the community and the actions
of the state—then the purposes of its controlling intelligence,
of its technostructure, become of the highest importance. They
are not confined by the market. They transcend the market,
use the market as an instrument and are the chariot to which
society, if not chained, is at least attached. That the modern
corporation deploys such power the neoclassical model, of
course, denies. That it is the reality we here see. Thus the importance of the purposes of the technostructure. The next
chapters look more closely at these. The chapters that follow
then examine the way that power is deployed in support of
these purposes.

These purposes, it will presently be seen, are directly in the
service of the interest, including the pecuniary interest, of the
technostructure. Avarice no longer serves, with an intermediate
eleemosynary bridge, the avarice of others. This should be reassuring. The contradiction of neoclassical theory disappears.
And in its adherence to the belief that power is used by people
to pursue their own accurately sensed self-interest this argument is wholly orthodox. Those who suggest that the modern
corporate management pursues not its own pecuniary interest
but the pecuniary interest of owners to which it is no longer
beholden must, in accordance with all traditional attitudes, assume the burden of proof. It is they who make selfless service
to others an essential ingredient of their system.

How Power Is Used:
The Protective Purposes

POWER is the ability of an individual or a group to impose its purposes on others. Its existence invites three questions: Who possesses the power (something that is not always evident); to what ends is it used; and what are the instruments that are employed in winning the consent or obedience of others? In the planning system, the economy of the large corporations, the power is possessed by the technostructure, and increasingly so with increased size and maturity. The instruments by which it is exercised are the subject of chapters that follow. It is necessary now to inquire as to the purposes this power serves.

In the most elementary sense these purposes are the same for all firms, large or small. The small entrepreneur seeks first of all to protect his position or authority—to stay in business and keep himself in his job. This, without excessive novelty, may be called his protective purpose. Then, having reasonably secured his existence, he seeks to advance his interests—to pursue his *affirmative* purposes. In the small firm both purposes involve earnings. The protective purpose is served by a certain minimum of return; if the entrepreneur does not have this, he loses his capital and therewith his right to continue in command of the enterprise. And his affirmative purpose, it is commonly assumed, is to make as much more than the minimum as may be possible without excessive risk, i.e., without jeopardizing too gravely the minimum return that serves his protective goal.

The purposes of the technostructure similarly divide into those that serve protective and affirmative interests. The purposes served are, as compared with the small entrepreneur's, more complex. This is because, having more power—not being at the mercy of the market in the manner of the small entrepreneur—the technostructure has latitude in selecting and pursuing its goals.

2

The technostructure has two protective needs. It must, like the small entrepreneur, secure its existence; it must prevent anyone —an adverse stockholder, an unpaid creditor—from throwing it out. And, short of this, it must minimize the danger of external interference with its decisions. As the last chapter has shown, all important decisions are taken collegially; only in this way can all relevant and specialized information and experience be brought to bear on the decision. And, we have seen, interference with these decisions by any individual who is not a party to the collegial process will be uninformed and most likely damaging, and will certainly seem so to those who, by their own participation, are aware of the range of information that has been brought to bear on the question. As the firm becomes larger and its decisions become more complex, it will increasingly be impelled to protect its decision-making from the uninformed intrusion of outsiders.

There are four possible sources of such intrusion. The first is owners and creditors. It is uniquely the right of the owner of the capital, in what is still called capitalism, to control the operations of the firm—the basic legal prerogative of the capitalist is to command. Equally the institution which lends the firm money has the right, at least before lending, to inquire how the loan is to be used. And it has a right to the assets of the firm in event of default. In between it has a greater or less right to protect its investment.

The three other possible sources of intervention are workers (ordinarily through their unions), consumers and the government. The technostructure of the mature corporation resists all external intervention. Far more tact and discretion will ordinarily be used in excluding the owner and creditor (and perhaps the consumer) than the union or the state. Indignation and oratory can be brought to bear against unions or the government. Stockholders are excluded by giving them the impression but not the substance of power. But the need—that of excluding the uninformed outsider—is the same.

The basic strategy by which the technostructure protects its decision-making process from owners or creditors consists in ensuring a certain minimum (though not necessarily a low) level of earnings. Nothing else is so important. Given some basic level of earnings, stockholders are quiescent. They become aroused, either individually or collectively, only when earnings are poor or there are losses and dividends are omitted. Proxy fights for the control of large firms, the case of takeover efforts apart, occur only in the case of those that are suffering low earnings or losing money. Among the hundred or so largest corporations (which account for the major share of all sales and assets) proxy battles, when earnings are good, are virtually unknown. This is another way of saying that the position of the technostructure under these circumstances is invulnerable.

Among the smaller of the large firms here considered—perhaps the smaller eight hundred of the largest thousand—low or interrupted earnings make the firm vulnerable, other circumstances being favorable, to a take-over bid. The earnings being low or absent, the stock is cheap; the stockholder is thus open to an offer by another corporation at some price above the going market. From the stock so assembled comes the power to dispense with the senior figures in the technostructure. Such senior members will not welcome the resulting threat to their tenure. The rest of the technostructure, being by its nature indispensable, will survive. But here too there will be a period of unease and uncertainty, for, with the change in control, there will be a tendency to study and rearrange organization. Also those involved in the take-over will often be unaware of the limitations of their own knowledge of the firm being taken over. In consequence the danger of uninformed intrusion on substantive decisions—ill-informed action on new or old product lines, ill-considered investment, uninformed acquisition or disposal of assets—is, at least for a time, very great. This is one reason why, in recent years, when firms with indifferent earnings have been taken over by a new and energetic conglomerate, they have frequently done worse.

Below a certain size there is no absolute defense against a take-over bid to the stockholders. But a favorable earnings performance remains the best protection. It is an indispensable part of any appeal to the stockholders to refuse bids for their stock.

A basic level of earnings also provides the firm, and therewith its technostructure, with a source of savings and thus of capital that is fully under its own control. This protects it from creditors. Not needing outside funds, it does not have to make any concessions to the views of those who would provide them. Nor, not having debt, does it risk their intrusion if earnings should fail. And, if it must borrow, it does so under conditions that protect its autonomy. The fact of earnings is the proof of competence in its decisions. A firm that is not making money must appeal out of weakness to bankers, investment bankers and insurance companies.[1] As a supplicant it cannot resist

---

[1] It will frequently have their representatives on its board of directors as a way of informing and reassuring the financial community and thus facilitating routine financing. Such directorships are also a minor form of business patronage. These directors, normally compliant and even somnambulant, will often come awake when earnings fail. See Peter C. Dooley, "The Interlocking Directorate," *The American Economic Review,* Vol. 59, No. 3 (June 1969), p. 314. Professor Dooley, although conceding substantial autonomy to the technostructure, believes that outside directors speak influentially for "the local

questions about, and possible interference with, top personnel or substantive decisions, for the absence of earnings encourages the presumption that external interference will improve performance. Like all other external influences this too will usually be uninformed and hence damaging. A firm that falls under the control of the bankers is quite likely to do worse. There is reason why this could be so.

Thus the first protective purpose of the technostructure. It is to ensure a basic and uninterrupted level of earnings. Whatever serves this purpose—the stabilization of prices, the control of costs, the management of consumer response, the control of public purchases, the neutralization of adverse tendencies in prices, costs or consumer behavior that cannot be controlled, the winning of government policies that stabilize demand or absorb undue risk—will be central to the efforts of the technostructure and corporation.

3

The technostructure of the firm in the planning system is an overhead cost—the outlay does not vary in close relation with sales or production. Members of an organization depend for their competence, in part, on their experience in working together in the organization. Such experience teaches participants who has information, how much each man's information is to be trusted. Accordingly members cannot, without damage, be hired and turned off with changes in operating levels in the manner of an old-fashioned labor force. It is necessary to *protect* the organization. Also the technostructure is itself the directing force. There is a special poignancy about cutting a payroll when the payroll consists, more or less, of those who are cutting it.

The counterpart of the technical character of production is

community" as well as on financial policy and the need to restrain damaging competition. I think, in showing the extent of interlocking directorates, he may be concerned to find them a greater source of influence than they are.

heavy investment in both fixed and working capital. This investment is also an overhead cost, although the fact that earnings are a source of capital—and do not involve an increase in fixed interest charges—does much to make it less dangerous. Finally, modern conglomerate growth, as will be noted presently, depends heavily on borrowed funds and thus adds to fixed overhead.

If demand and prices fluctuate and costs are given, earnings obviously are vulnerable. The firm in the planning system has, therefore, a further and powerful inducement to control those factors—prices, costs, demand, actions by the state—which secure its earnings. Exercise of power in response to the protective motive should not be thought of as a willful thing. It is largely compelled. Technology and associated organization and capital requirements lead the firm, in the interests of its own survival, to impose its needs upon the society—to control those forces in its environment which might threaten its earnings.

The effort of the technostructure in securing a minimum level of earnings is not completely uniform or successful. Conglomerates, as the next chapter shows, have emphasized growth in preference to security of return. In consequence their earnings have been less reliable than those of other large firms. In the early seventies some weapons firms had to contend with managerial tasks which, at least temporarily, were beyond their capacity. Nonetheless, among the largest corporations, the certainty of earnings is very great. In 1970, a generally poor year for corporate earnings,[2] only six among the hundred largest industrial firms failed to return a profit. These, with the single exception of the Chrysler Corporation, were either weapons firms or conglomerates or a combination of the two. Among the largest financial and merchandising corporations losses were even more exceptional. In 1971, seven among the hundred largest industrials lost money. Again, with two exceptions,

[2] See Chapter XVII.

they were weapons firms or conglomerates. Only two of the same companies lost money in both years.[3]

4

As noted, the protective purposes of the technostructure also require the exclusion of unions, consumers and the government from interference with collegial decisions. Here traditional economic attitudes, powerfully abetted by convention, are the basic defense.

In the neoclassical model the firm is ultimately subordinate to the market and thus to the consumer. There is no general reason why the consumer (or the government acting on behalf of the consumer) should interfere with the decisions of the firm. The consumer is already in control. Fraud and deliberate befuddlement of the consumer must be prevented. But as long as the consumer is not misled as to his interest, the system will conform to that interest. One of the instrumental and very useful services of neoclassical economics to the planning system is in leaving those who are exposed to its instruction with the impression, however vague and undefined, that interference with private business decision is unnecessary and abnormal.

The government is excluded on similar grounds. The public through the consumer being already in charge, the public through the government need not and should not intervene. This doctrine, reinforced by convention so nearly unchallenged as to be largely unrecognized, forbids government interference with any managerial decision of a *private* corporation. The selection and design of its products, the method of their production, the selection, promotion or pay of personnel, all other such matters, are *private* business. Even where there is a powerful public effect from such decisions, as in the safety of the

[3] Several railroads also lost money. This, however, if anything, sustains the present case. Government regulation has generally kept the railroads from gaining control of the factors influencing their earnings. And, partly in consequence perhaps, they have a special tradition of feckless performance.

design of automobiles, the environmental effect of detergents, the dissonance of radio commercials, the encouragement to crime and violence in television programming or (until recently) the promotion of cigarettes as an aid to health, a heavy burden of proof lies on any intervention. It may never be in relation to a particular business decision; interference must always be by general rule. And to say that something "is for management to decide" is ordinarily to end all argument over public intervention.

Needless to say, incompetence is not an excuse for intervention in the business firm. As some men are mediocre, so are some organizations. And there is a plausible tendency for the mediocre organization to perpetuate itself—a modestly inadequate man will seem a genius among men of more determined mediocrity. He will win promotion and extend his mediocrity over a larger area of responsibility. And his success will often be welcomed by his colleagues, for, in contrast with a man of ability, he will be more tolerant of stupidity.

None of this may be a matter of public concern even at the highest level of development where adverse consequences from the market are fully excluded. Before World War II, the Ford Motor Company had, for a number of years, assiduously cultivated incompetence. As a result its performance in building the B-24 bomber was disastrous. The great factory was built at Willow Run; the aircraft, though much needed, did not for a long time emerge. The discussion in Washington among the businessmen concerned with war production was intense. That the Ford management was appalling was agreed. But to breach the principle of managerial autonomy was thought impossible —even for winning a war. And, to the relief of everyone, the principle was preserved. After many months the airplanes began to come out. In more recent times there was a widespread belief, not ungrounded in evidence, that the management of the Lockheed Corporation was inadequate in a uniquely expensive way. Although almost all of its orders came from the government and the government was guaranteeing its debt, the right of the corporation to run its affairs in its own way went largely unchallenged.

5

Interference with managerial decision by workers, notably by unions, is similarly excluded. The firm in the neoclassical system seeks the combination of labor and capital which minimizes cost at any relevant output, and thus it maximizes its earnings. Any interference with the decisions that produce this result will raise costs. Costs being higher, prices will be higher, and consumption and production and employment will be lower. Thus the interference is ultimately damaging to labor. The accepted economics and the resulting attitudes thus hold it to be in the interest of labor itself to eschew any interference with managerial decision.

Some unions do intervene on decisions involving mechanization and the related use of manpower. This, however, is almost universally disapproved. A management that is thought to be technologically backward is believed to damage only itself, but a union that resists innovation is socially reprehensible. By the same token unions that accept all innovation, whatever the effect on employment, have a high standing in the convenient social virtue. As in the case of the United Mine Workers in the recent past this may be sufficient to convert leaders of the highest level of rascality into labor statesmen.

Thus the protective purposes of the technostructure and how they are served. However distinct conceptually they are commingled in everyday decision with the affirmative purposes to which we now turn.

XI

The Affirmative Purposes

THE PRIMARY AFFIRMATIVE PURPOSE of the technostructure is the growth of the firm. Such growth then becomes a major purpose of the planning system and, in consequence, of the society in which the large firm is dominant.

The first effect of growth is to reinforce the protective purposes of the technostructure. The large firm, as we shall presently see, is, with exceptions, better able than a small one to control its prices and costs; to persuade and manage its consumers; for the foregoing reasons, to protect itself from competitive reduction in profits and thus to protect its earnings and therewith its source of capital; to pass on labor costs that it cannot control; and to win needed attitudes in the community and action by the state. This means it can better protect itself from the adverse movements in earnings that might invite the interference of stockholders or creditors and from the adverse public attitudes that might invite the intervention of unions or consumers or the government.

But the growth of the firm also serves as does nothing else the direct pecuniary interest of the technostructure. In a firm that is static in size an individual's advancement awaits the death, disability or retirement of those above him in the hierarchy. Or it depends on his ability to displace them. And, as he may hope to displace others, others below will hope to displace him. As he must succeed in aggression against others, so others will be moved to aggression against him. As he will be regarding his seniors, however discreetly, for promising indications of prospective disability or morbidity, so others with similar hope will be watching him.

In a growing firm, in contrast, new jobs are created by expansion. Promotion ceases to be a zero sum game in which what one wins, another loses. All can advance. All can succeed. A cooperative working relationship is not impaired by reciprocal hopes for alcoholism or an automobile accident. And as sales, number of employees or the value of assets managed increase, so do salary, expense accounts and the individual's claim to nonsalary income or privilege. The scale of his office and the excellence of its furnishings are enhanced; his access to private lavatory and company plane are assured. So is his reward from employee obeisance and peer group homage.[1]

Growth also gains in importance as a goal because of the close relationship between those responsible for it and the resulting reward. This is a most important and much neglected matter. In a large organization earnings are calculated for divisions or "profit centers" of substantial size. The contribution of any subordinate individual or group to earnings is merged with that of many others. Under all circumstances it is subjective—a matter for discussion and estimate.

[1] A goal that could, of course, be served by moving to another, larger firm, as not infrequently happens. Robin Marris of the University of Cambridge, who has given these matters the most detailed thought, has dealt with this point—and the motivation to growth in general—by suggesting that "were managers perfectly mobile, they could move progressively from smaller to larger firms, finally crowning ambition by becoming Secretary of Defense. But this is not, in fact, the way things go. On account of a general preference for internal promotion, mobility is [relatively] low and ceteris paribus a man who knows and is known to the firm is of considerably greater economic value than a comparable outsider. Therefore managements are likely to see the growth of their own organization as one of the best methods for satisfying personal needs and ambitions, an attitude which is reinforced by psychological tendencies to identify the ego with the org. In oligarchic firms, where policy is determined by groups, the policies most likely to command agreement will be those expected to increase the utility of every member. And if we think of utility as a basket of salary, power and prestige, it is clear that for those at the top, growth is an outstanding candidate, even though some of the increment may have to be shared with newly promoted juniors." "A Model of the 'Managerial' Enterprise," *The Quarterly Journal of Economics,* Vol. 77, No. 2 (May 1963), pp. 187–188.

In the case of growth, in contrast, the contribution of the individual or a small group is direct and visible. The sales figures of a new product, gadget or line are a datum even if their contribution to earnings is less clear; and those responsible are known. Indeed growth often rewards directly those who are responsible for it. The unit, however small, that expands its sales therewith expands its employment and the claim of those responsible to the promotion, pay and perquisites that go with the larger operation. The engineer who sees some hitherto undisclosed opportunity for product development will have his responsibilities, and therewith his position and pay, enhanced by that development. The marketing man who successfully persuades the public to buy some abnormally improbable artifact will, in consequence, be in charge of the resulting larger marketing operation. He promotes himself along with the product. The possibility of such self-reward is widely diffused through the technostructure. Many of its members having a direct stake in growth, it is not surprising that the whole corpus of the technostructure is deeply committed to growth.

2

If the economic system as a whole is growing, firms in general will be expanding. To the opportunities for promotion within the firm will be added the opportunities from other firms. With so many influential people finding the growth of the firm and the associated growth of the economy to their advantage, it would be surprising if they were not to conclude that economic growth is a good thing. They do. In consequence, or largely in consequence, economic growth has acquired the highest standing as a social goal. What contributes to the growth of the economy, and therewith to the pecuniary well-being of the technostructure, is strongly blessed by the convenient social virtue and is praised on all occasions of public ceremony.

Economists have generally credited themselves with a special awareness of the social advantages of economic growth. They saw that it would mean more consumption, more income, more employment, greater tax revenue, more social services, greater happiness. It was a convenient buttress to the economist's enlightenment that such growth also served the affirmative interests of the technostructure. Had its contribution been adverse, growth would not have been well regarded.

3

The growth of the modern corporation is a complex process involving a blend of strategies. The first strategy is the expansion of production and sales by the existing corporate entity. This, it is commonly assumed, involves adding products or services which have some technical complementarity—which, in the manner of the rubber manufacturer going into plastics, uses some of the same technical expertise, plant and equipment, marketing outlets, marketing knowledge or managerial skills. But where growth is the goal, such complementarity is not essential. The decisive question is what contributes to growth, and this may come equally or in greater measure from wholly unrelated products. In recent times there has been much puzzlement over the evident illogic of the mix of goods produced by the modern corporation. This should not be a matter of surprise; its primary concern is not with technical complementarity or whether similar markets are served. Its primary concern is with growth.

The second strategy is the acquisition of smaller firms in related or unrelated areas for more or less complete consolidation with the technostructure of the acquiring firm. This allows the superior financial resources of the larger firm to be used for a far more rapid growth than would generally be possible from the expansion of sales. Such acquisition is also exceedingly valuable for contending with superior managerial, technical or marketing competence. These may, on occasion, allow the smaller firm to make gains in growth in comparison with the larger and, also on occasion, more conservative or bureaucratic rival. But it is ordinarily open to the latter to protect itself—sooner or later—by using its greater financial power to purchase the small but inconvenient rival.

The full development of this second strategy is agglomeration. This involves the acquisition of voting control of one corporation by another with the purpose, however, of leaving the former's technostructure and therewith its operating autonomy at least partially undisturbed. Not infrequently it involves acquisition of financial enterprises, such as insurance companies, which have a large flow of cash for investment. This can then be placed at least partially in the service of acquisition of other firms. Such agglomeration is the most rapid form of expansion; in one of the more spectacular recent examples it propelled International Telephone & Telegraph from forty-seventh place among industrial corporations in 1961 to ninth place ten years later. Since the firms that comprise the conglomerate operate in different industries serving very often radically different markets, not much is gained from growth in control of prices, costs or capacity to manage or persuade the consumer. The principal advantage, as regards power, is in the influence that can be deployed in relation to the government and in the possibility of using (or milking) the capital resources of one unit for the needs of another. Much of the drive to agglomeration reflects the interest in size qua size—it is the purest example of the effort of the corporate leadership to reach for the rewards of bigness for itself.[2]

---

[2] Above a certain size the coordination of functional departments of a firm— purchasing, product development, production, marketing—imposes an unduly heavy burden of coordination on the top management. The solution is to create divisions with full authority over their own procurement, production, sales, and with companion responsibility for profits. The conglomerate in which each unit is a separate corporate entity is the most developed form of this structure. Some have urged that this structure tends to rehabilitate profit maximization as a corporate goal since profits are the prime test of divisional performance. That this structure helps to protect profits—that it serves the

While the strategy for the most rapid growth, agglomeration is also the most uncertain and the one that is in most serious conflict with the protective purposes of the technostructure. The uncertainty arises from the possibility that the technostructure of the firm being acquired will successfully resist the take-over; the larger it is and the more jealous of its resulting autonomy, the more likely such resistance. And if the firm being taken over is large in relation to the acquiring firm, outside borrowing will be required. This increases fixed charges and adds to the vulnerability of earnings. And even if the organization is financed by an exchange of stock, this may put power in the hands of new and intrusive owners.

In the late nineteen-sixties, in the so-called conglomerate explosion, there was a large amount of this high-risk growth. By no means all reflected the initiative of large firms with highly developed technostructures. A considerable part was on the initiative of individual entrepreneurs operating from a base in a firm of considerably less than the largest size. Typically they combined indifference to, or unawareness of, risk with a distressing capacity to impress others with their estimate of their own financial acumen. Most of the resulting structures did badly; several were subsequently in severe financial difficulties. It is possible that this high-risk, individually sponsored agglomeration was an aberrant development associated with the then-current boom psychology.

At all times, however, growth by acquisition is a wholly normal tendency of the planning system. Between 1948 and 1965—years that exclude the frantic mergers of the latter sixties—the 200 largest manufacturing corporations in the United States acquired 2692 other firms with total assets of

---

protective purposes of the technostructure—is not in doubt. But the volume of sales is also a central test of divisional performance. And those who make the case for the effect of this structure on profit maximization themselves concede that its primary purpose is to remove the limits on growth and make almost any size of firm possible. See Oliver E. Williamson, *Corporate Control and Business Behavior* (Englewood Cliffs, New Jersey: Prentice-Hall, 1970).

$21.5 billion. These acquisitions accounted for about one seventh of all growth in assets by these firms during this period. Excluding the 20 largest manufacturing firms, they accounted for between one fifth and one fourth of the growth.[3] During the next three years the 200 largest corporations acquired some 1200 more firms with additional assets of approximately $30 billion.[4] The largest 200 are now estimated[5] to control nearly two thirds of the assets of all companies engaged in manufacturing.

It may be urged, once more, that to be successful the acquisition of one company by another need not increase the rate of return on the capital of the combined enterprises. This is not the primary purpose. The primary purpose is to enhance the pecuniary reward and prestige of the technostructure of the acquiring (and in some cases also of the acquired) firm. This is accomplished not by a greater rate of return but by greater size.

At first glance there is a superficial parallel between the growth of the modern large firm by acquisition and the process of capitalist concentration as adumbrated by Marx in which the smaller capitalists are devoured progressively by the large. The comparison is not precise. The motivation to the Marxian process was exploitation—and profit. The primary motivation in the modern process is bureaucratic advantage—the enhanced prestige and return of the technostructure. In this process the power of the capitalist is, if anything, diminished. It is that of the technostructure which is enhanced.

A firm that seeks to maximize its profits, as in the neoclassical model, must be large enough to use the most efficient size

[3] William G. Shepherd, *Market Power and Economic Welfare* (New York: Random House, 1970), p. 75. Data are from the testimony of Willard F. Mueller, Hearing before the Select Committee on Small Business, United States Senate, 90th Congress, 1st Session, March 15, 1967, p. 455.

[4] Morton Mintz and Jerry S. Cohen, *America, Inc.* (New York: Dial Press, 1971), p. 36. These totals include the acquisition by very large firms of smaller ones within the largest 200.

[5] By Willard F. Mueller of the University of Wisconsin.

of plant. And it will have further gains from size if this allows it to control its prices and thus the gains from monopoly power. It should not be larger; if it is, it sacrifices efficiency and therewith profit. "[There] are undisputed advantages to large-scale integrated operations at a single steel plant . . . but there is little technological justification for combining these functionally separate plants into a single administrative unit. United States Steel is nothing more than several Inland Steels strewn about the country . . . A firm producing such divergent lines as rubber boots, chain saws, motorboats, and chicken feed may be seeking conglomerate size and power, it is certainly not responding to technological necessity."[6]

Excessive size is commonly explained in the orthodox pedagogy by irrationality—the commitment of the great corporation to a mindless giantism which serves no purpose. The very great size of the modern corporation is a highly important fact. To attribute it to stupidity will not, to many, seem intellectually satisfying. But when the purposes are seen to be those of the technostructure, there is no problem. Size and growth serve its purposes without limit. They serve the protective purposes of the technostructure and the bigger the firm, the better, in general, the protection. And the greater the growth, the greater the tangible pecuniary and other reward to the technostructure. One test of economic ideas is whether they fit together or must be jammed together. The present view of the purposes of the large corporation is, by this test, reassuring.

4

The two main purposes for which the technostructure uses its power are now in view. The technostructure protects the autonomy of its decision-making primarily by seeking to se-

[6] Walter Adams. Hearing before the Select Committee on Small Business, United States Senate, 90th Congress, 1st Session, June 29, 1967, pp. 12–13.

cure a minimum level of earnings. Then it rewards itself affirmatively with growth. This, however, is not all. Where the firm has a strong technical orientation, technical development and innovation may have a limited life of their own. They will be pursued, though within narrow boundaries, for their own sake. This is a matter to which I will return. And partly for reasons of tradition and more because of the objectivity of the test, the firm will ordinarily seek to show an annual improvement in its earnings.

Profits were the purpose of the capitalist firm. They are still emphasized to the near-exclusion of all other purposes in neoclassical pedagogy. Stockholders and their spokesmen in the financial community obviously emphasize their importance. As a practical matter no one either inside or outside a company can tell whether profits are at a maximum. And there is no agreement as to the period over which profits are to be maximized. Since costs can be postponed and since consumers do not react immediately to prices that are out of line with those of rival firms, it is often possible to have higher profits in the short run at the expense of the long. Or—a common case with supermarkets, department stores and other merchandising firms—lower margins and profits in the short run may mean higher volume and higher returns in the longer run.

Although no one can tell whether profits are at a maximum, it is possible to know whether they are increasing or decreasing. And if accounting and depreciation practices remain unchanged, this test—like an increase in sales—is objective. So the trend in earnings is also a test of performance. A growing firm with good earnings is doing well. A growing firm with increasing earnings is doing better. Thus to the other purposes of the technostructure may be added the effort to show an improvement in profits from one year to the next. It does not serve the interests of the technostructure as directly as does growth. It is nonetheless an important goal.

5

The several protective and affirmative purposes of the technostructure are not necessarily consistent, one with the other. While growth, generally speaking, adds to the power of the technostructure and thus reinforces its ability to ensure a minimum level of earnings and so serves its protective purposes, some kinds of growth, as we have seen, involve increased risk. The need to show an improvement in earnings may conflict with the goal of growth. Technical development may be in conflict with security of earnings.

In the neoclassical firm there was one goal—the maximization of profits. In consequence there was one pattern of behavior—one theory of the firm. Scholars who have admitted to the possibility of other goals than profit maximization—who have conceded that the firm might seek, for example, some combination of security of return and growth—have continued, nonetheless, to seek a single explanation of how the firm behaves.[7] This is a serious error. There is no reason to suppose that the technostructures of firms in different industries will make the same compromise between inconsistent purposes. An advanced electronic, chemical or computer firm, densely populated with engineers and scientists, will set more store by technical innovation as a goal than a meat packer or a steel company. In other cases there will be differing emphasis on security of earnings as opposed to growth. The choice of the largest firms on these matters will differ from that of the less large. And differing concessions will be made to the need to show improvement in earnings. The social consequences of these choices, as well as of the power with which they are pursued, can—as we shall see—be considerable.

This is not a result that should at all surprise us. We ex-

[7] See Robin Marris, *The Economic Theory of "Managerial" Capitalism* (New York: Basic Books, Inc., 1968) and my foreword to this edition.

pect public bureaucracies—the Pentagon as opposed to the Department of Labor or the Department of State—to pursue different goals in consequence of different size, power and security of position. This is the nature of organization—or, more generally, of social performance under planning. The remarkable thing, rather, is that anyone should imagine that the purposes and thus the behavior of American Telephone and Telegraph, General Motors, LTV, Corning Glass, Control Data, Seagram's, not to mention Volkswagen, Renault and Mitsubishi will all be the same. No theory should lead to totally implausible conclusions.

XII

How Prices Are Set

IN THE NEOCLASSICAL MODEL prices are primary; they are the intelligence network of the economic system. They signal changes in the wants of the consumers to the producing firms. And, in reverse, they tell the consumer of changes in the costs of production and, therewith, of new opportunities for serving his or her needs. On the basis of the prices thus established the consumer so distributes purchases as to maximize the satisfaction derived from the funds available for expenditure. And since nothing is so important as the consumption of goods, the consumer thus maximizes happiness. The agreeable equilibrium so established has its counterpart in the way labor, capital, raw materials and managerial ability are distributed. Prices, including that of labor, are also the signal to these factors of production as to where they can most profitably be used. Their ultimate deployment also reflects the will of the consumer. This use, monopoly and some minor hindrances apart, is, from the viewpoint of consumers (and given the distribution of income), the best that is possible.

It is through prices that the neoclassical monopoly or oligopoly exploits the power that goes with being one, or one of the few, sellers in the market. Such power allows of prices and profits that are higher and output that is smaller than would be the case were sellers more numerous. In consequence consumers pay more and have less product or service than is necessary or desirable. And smaller amounts of labor, capital and materials are committed to the product or service than would be ideal. And more workers must find employment elsewhere.

And the distribution of income is distorted in favor of the monopolist. Prices, in the neoclassical model, are thus the prime clue both to the perfections and to the imperfections of the economic system. Not surprisingly the way prices are established is a major preoccupation of neoclassical economics. Until comparatively recent times the study of economics consisted of learning how prices and incomes were set, and not much else.

In the market system—in the real world of the small firms that are barred from the planning system by their inability to use organization—the role of prices is less pure. There is an admixture of monopoly, competition and—as in the case of agriculture—government regulation. The very different world of the planning system is also adjacent—with powerful effect on resource distribution. Still the small manufacturer, retailer or service enterprise has power to control prices only within narrow limits. And the price which the United States Department of Agriculture establishes for wheat or corn is beyond the influence of any individual farmer. Accordingly prices remain a datum external to the firm. It must accommodate its production to what it cannot control. In inducing entry into, expansion, contraction or abandonment of, the business, prices still guide, however imperfectly, the distribution of resources as between products and services. So prices in the market system remain important.

In the planning system the role of prices is greatly diminished. They are much more effectively under the control of the firm. And they are only one—although still the most visible —of the forces which are beyond the influence of the firm in the neoclassical model or the market system but which are subject to its influence in the planning system. In the market system consumer behavior, costs, the response of suppliers, the behavior of the state are all beyond the reach of the individual firm. In the planning system the firm seeks and wins power or influence over all of these things.

It follows that prices are no longer of unique importance in telling how resources are distributed. What counts is the whole deployment of power—over prices, costs, consumers, suppliers, the government. Prices may be less important than the energy, guile or resourcefulness with which the firm persuades the consumer or the government to want what it produces or by which it eliminates the possibility of choice. The level of costs may be less important than the energy with which the firm plans its procurement. What it now produces is the result of its past skill in winning government support for its research and development with resulting processes or products. The distribution of resources in the planning system, a most important point, is the result not of the firm's control over prices but of the aggregate exercise of all its powers.

In the planning system, then, the use of resources no longer reflects, by way of prices, the decisions of consumers. The notion of consumer purchases so distributed as to win a resource use that maximizes satisfactions dissolves in irrelevance. The distribution of resources extensively reflects the power of the particular firm, in company with other firms, to pursue its own purposes, and it is this power that allows us to speak of a planning system. A preoccupation with prices, it will be evident, could conceal the more general exercise of power by the modern firm. It could divert attention from the more comprehensive exercise of power—the planning—of which the control of prices is only one part.

2

The control of prices by the firm in the planning system, like the other uses of its power, is governed by the protective and affirmative purposes of the technostructure. That is why these purposes must be fully in view.

The first protective requirement is that prices be under tight control. This keeps adventitious and adverse movements in prices from damaging or destroying earnings. In the market system technology is simple, and capital, accordingly, is comparatively unspecialized. As with multiple-purpose machine tools or general-purpose factory structures such capital can be shifted from one use to another. Additionally the period between initiation and completion of production is short. This means that if prices become unfavorable, the entrepreneur can turn promptly, though not always without pain, to the production of something else. In the planning system, in contrast, technology is complex, and it is associated with capital equipment —machinery and plant—that is specialized to a particular service or product. The production period—the time that elapses between a decision to produce and the emergence of the product—is much longer than in the market system. Capital equipment, being specialized to a particular product, must be designed as well as produced. And on occasion the equipment to make the equipment must be designed. Finally, with the increasing importance of organization, there is the cost of the technostructure itself. So, in the firm in the planning system, heavy costs are incurred before there is a salable product, and these continue regardless of the return from sales. In such circumstances prices and costs must be under control. So also, to the extent possible, consumer and government demand. So also supplies at the controlled costs. Planning, as previously noted, is not a willful act of the large enterprise; it is inherent in the whole matrix of development of which advanced technology, intensive use of capital, the rise of the technostructure are a part.

Prices must also be controlled because some costs, notably those of labor, are not fully within the power of the firm. To protect itself the firm must be able to increase prices in order to offset increases in labor costs that it cannot prevent. This, in its inflationary consequences, is a matter of no slight practical importance. The firm must also control prices and customer and supplier response because technological development tends to make demand increasingly inelastic and markets increasingly erratic. An increase in the supply of asparagus or carrots coming onto the market will reduce prices and expand consumption in a reasonably predictable and orderly way. An increased output of commercial aircraft dumped similarly on the market would have highly unpredictable effects on prices. To release an increased supply of high-capacity computers for what they would bring would be unthinkable. The situation is similar with labor, materials and components. The market is a reliable source of unskilled labor which can always be had at or a little above the going wage. Specialized engineering talent is not similarly available on short notice in response to higher wages. Nor are esoteric materials or components. Instead of relying on higher wages or prices to procure labor, materials or components the firm must set wages, salaries and prices and then concentrate its energies on getting the requisite supply at these levels and prices.

3

The control of prices in the planning system is not a matter of great difficulty. Rather it is the automatic consequence of the development in this part of the economy. Firms that are large are large in the markets they supply.[1] What these firms produce and sell affects their prices, which is to say they have power over their prices. When a firm has acquired this power, it is the price rather than the intended production which, ordinarily, it first establishes. This is worth a word. The grain or livestock farmer is small in his market, and it makes no difference to the price whether he increases output, decreases output or leaves the business. His only choice is to take the price as given and adjust his production accordingly. General Motors or the United States Steel Corporation, by contrast, would drastically affect the prices of their cars or steel by dou-

[1] A neoclassical fiction holds that there is no *necessary* relation between firm size and market power and hence no relation at all. The point is not worth arguing. The large firms of the planning system are also relatively large in their markets. And this is so virtually without exception.

bling output and putting it on the market. Their safest recourse is to exercise the initial control not over production with its uncertain effect on price but to fix the price. Production is then accommodated to what can be sold at that price.

The power to set the price means that any other major firm in the industry—Ford or Chrysler in the case of automobiles or Bethlehem or Inland in the case of steel—can, by fixing a lower price, force an alteration in the level first established. This may happen. But there is also a general recognition that such action, should it lead to further and retributive action by the firm originally establishing the price, could lead to general price-cutting. This would mean a general loss of control—a general sacrifice of the protective purposes of all the technostructures involved. The danger is recognized by all. In the planning system there is, accordingly, a convention that outlaws such behavior. It is almost perfectly enforced. No contract, no penalties and usually no communication are involved. There is only an acute recognition of the disadvantage of such competitive and retributive action for all participants. It is a remarkable example of the capacity, where money is involved, to recognize and respect a community of interest.[2]

4

The control of prices is for the purpose of securing a minimum level of earnings—for ensuring that uncontrolled prices do not

[2] The convention fails only where especially obdurate circumstances prevent the tacit acceptance of a particular price. This has happened in the case of heavy electrical products which, being produced to specification and sold under sealed bid, had no known and visible price. Or it has happened in the building supply industry where numerous producers or numerous and varied products made it difficult to enforce the convention by tacit means. The result has been overt collusion in these industries and spectacular antitrust prosecution. This has been taken to mean that these industries are especially wicked in their willingness to conspire to set prices. What it actually means is that tacit collusion in these industries is especially difficult. Accordingly the participants have resorted to illegal means to enforce precisely the same understanding on prices that all other industries where there are a few large participants are able to accomplish quite legally and, indeed, with an aspect of superior virtue.

plunge the firm (and industry) into loss. When this control is secure, the protective purposes of the technostructure give way to its affirmative purposes. These determine the level at which the controlled prices are set.

Growth is the most important affirmative purpose of the technostructure. Accordingly prices will be set at the level which, while ensuring the necessary level of earnings, has the primary effect of sustaining and expanding sales. The word primary must be emphasized; the level of prices which would reconcile maximum sales with the necessary minimum of earnings may be further adjusted by the need to show an increase in earnings. And, as noted in the last chapter, no single generalization as to the resulting compromise will be valid for all firms of the planning system.

The prices that are so set—that reflect the affirmative purposes of the technostructure—will almost always be lower, and on occasion much lower, than those that would maximize profits over some period relevant to managerial calculation.[3] The buyer cannot, at one and the same time, be attracted by prices that expand sales and repelled by those that maximize profits.

Additionally economics, as always, is a tapestry in which each part must be in harmony with the rest. The technostructure not only controls prices; it seeks to ensure the response of consumers at these prices. Prices must be so set that they are consistent with the need to persuade the consumer. The consumer cannot be persuaded if he is subject to monopolistic exploitation or if he believes himself subject to such exploitation. Finally, a point to be emphasized presently, the price that will be established in any industry will tend to be that which reflects the preference of the technostructure that is most strongly committed to growth. Its price will be lowest. Others must accept that price and therewith that goal.

[3] As earlier noted, there is often a choice between profits in the short run and profits based on larger volume in the more distant future. The conflict between profits and growth thus diminishes, the longer the period of calculation.

5

The process by which prices that serve the affirmative purposes
of the technostructure are established in an industry, like that
which establishes the basic control of prices, derives from the
frequently common interests of all firms and the shared sense
of what serves those interests. Different technostructures have
a broadly similar interest in growth—the prices that serve the
growth of one firm will, generally speaking, serve the growth
of others. One firm, accordingly, can set prices knowing that
others will find them broadly acceptable. And, the number of
firms in the industry being small, the pacesetting firm can know
or judge what will be acceptable to the other firms and be so
guided. The others will then conform. On occasion another
firm may express its preferred price, reflecting its different cost
position or its different compromise as between the affirmative
purposes (i.e., growth and increasing earnings) of its technostructure. The initiating firm will often then accept this higher
or lower price. The common protective purposes require such
cooperation. The resulting compromise will not be ideal for
any given firm but will be tolerable for all. Ensuring such compromise is the overriding recognition by all the technostructures
that independent action—action that does not take account of
interdependence—will have a worse result and jeopardize the
protective purposes of all.

Prices in an industry, so far as they can be compared, must
be roughly uniform. Small firms may, on occasion, take the
initiative in changing prices; such price leadership by an unimportant participant was once considered to have a settling effect
on the Department of Justice. But the position of the larger
firms is decisive. It is the General Motors decision on the price
for a standard vehicle that is ultimately controlling, not that
of American Motors. Similarly as between IBM and its smaller
rivals, between General Electric and Westinghouse and the
smaller manufacturers of electrical goods, between the oil ma-

jors and the independents. Since the large firms have the most fully developed technostructures, it follows that it is the purposes of the developed technostructures that are most strongly reflected in the pricing of the industry as a whole. These firms are most completely removed from the influence of the capitalist entrepreneur; they reflect, in purest form, the desire to combine a secure level of earnings with the growth that most rewards the technostructure. It is this purpose, other things equal, that is enforced on the industry as a whole.[4]

Having established a generally satisfactory level of prices, the practice in most industries is to leave them unchanged for an appreciable period of time. This reflects the generally imprecise character of the process by which prices are set. There is no exactly right level for all firms; one approximation is nearly as good as another; it is sensible to live with the approximation that exists. This administrative rigidity reflects, also, the protective purposes of the technostructure. Such small danger as there may be of loss of control comes when prices are changed. Then misunderstanding of another firm's motives in cutting a price or refusing an increase opens the way, however slightly, to competitive price-cutting. To minimize the number of price changes is thus to minimize this risk.

6

When prices are fully under control, they can be raised to offset increases in costs that are not similarly under control. Wages are the most important example of an uncontrolled cost. The ability to offset wage increases with price increases is of great importance to the protective purposes of the technostructure.

[4] Which limits the value of empirical studies of corporate motivation. Although these studies tend to show that the developed technostructures are more strongly oriented to growth than those in less mature corporations, this difference is obviously disguised if the larger firms are able to force their purposes on the smaller ones. See R. Joseph Monsen, Jr., and Anthony Downs, "A Theory of Large Managerial Firms," *The Journal of Political Economy*, Vol. 73, No. 3 (June 1965), p. 221.

In the market system the absence of control over prices means that the firm must, initially at least, absorb some or all of a wage increase. In the planning system, in contrast, increased wage costs can readily be passed on to the public. That this can be done—that any conflict with labor can be resolved at the expense of third parties—means, more than incidentally, a greatly diminished tension between workers and the technostructure within the planning system. It means also that a general price increase can be expected to follow a general wage increase, and this, in fact, is all but universally the expectation. Usually the price increase will be more than sufficient to offset the cost of the wage increase; this is because the occasion of the price increase following the wage increase is used also to rectify the level of earnings in favor of the firm.

That price increases usually follow wage negotiations shows, more than incidentally, that profit maximization is not a purpose of the technostructure. If revenues can be increased just after a wage increase, they could, obviously, have been increased well before.[5]

The elimination of price competition as part of the protective purposes of the technostructure does not similarly eliminate other forms of rivalry between firms. Competition in product development, advertising, salesmanship and public ingratiation continues. Unlike price competition the effects of such rivalry are limited. Price-cutting can plunge all firms into disastrous loss. Other forms of rivalry, although they can be expensive, have no similar potential for damage. On the contrary—as will be stressed later—each firm's selling efforts do something to sell

[5] A purely formal case can be made that the change in marginal cost with the wage increase alters upward the price at which profits are maximized. The point is not of practical importance. A better case could be made that profits, were they being maximized, would only be so maximized over a much longer span of time. Accordingly one should not be surprised if the adjustment at any given moment is imperfect and is perfected in the wake of a major cost change. I raise the point only because the more devout defenders of the neoclassical model set store by showing that a price increase following a wage increase can be explained by marginal cost-price relationships. They do not, however, make the most plausible case even in terms of the neoclassical model.

the products or services of the industry as a whole and to affirm the happiness which derives from consumption in general. In price competition there is damage for all; in other forms of competition there can be benefit for all.

7

The commitment of the technostructure, then, is to a price policy that first serves its protective interest in ensuring that prices be under control and then serves its affirmative interest in growth of the firm (with its rewards to the technostructure) as modified by the need to show improving earnings. This view of prices, in turn, illuminates one of the major conflicts between the conclusions given by the neoclassical model and the reality. In the typical industrial market—that of few firms or oligopoly —prices in the neoclassical model are held to be set so as to reflect the maximum return to producers as a group. This, subject to some imperfection in the tacit communication between oligopolists, is the same price that would be charged by a monopolist. No point is better accepted by the neoclassical model than that the monopoly price is higher and the output smaller than is socially ideal. The public is the victim. Because of such exploitation, oligopoly is wicked.

Yet exploitation by modern oligopoly leads to no serious public outcry that production is too small or prices too high. The automobile industry, rubber industry, oil industry, soap industry, processed food industry, tobacco industry and intoxicants industry all fit precisely the pattern of oligopoly. All are held by neoclassical theory to maximize profits as would a monopoly. In all, comparative overdevelopment—as compared, for example, with housing, health care, urban transit—is regularly cited in complaint. Or the effects of their growth on air, water, countryside, health are held against them. Never—literally—is it suggested that their output is too small. Nor are their prices a major object of complaint.

We now see the reason—and we begin to see one of the major dividends from a clear view of economic reality. The firms in these industries control prices in response to protective need —in response to the heavy capital investment, long-time horizons, extensive specialization and organization and, in consequence, the higher proportion of overhead costs. This same technology and organization allows of increasing productivity and falling costs. Of these the public approves. These firms set prices with a view to expanding sales—to growth. Of this, the antithesis of monopoly pricing, the public also approves. The neoclassical model describes an ill that does not exist because it assumes a purpose that is not pursued. And proof lies in the fact that the ill it describes provokes no grave public complaint. It is inconceivable that the public could be universally exploited without being aware of it.

Yet, as usual, the neoclassical view renders service to the planning system. Monopoly is a word with an exceptionally unsavory connotation. An especially talented burglar may command admiration but not a monopolist. While exploitation of the consumer is not a problem, there are other grave problems arising from the exercise of power by the planning system. As long as this is associated with monopoly, a single, harmless remedy comes automatically to the properly educated mind, which is to break up the offending firms—to eliminate the monopoly power by having several or many small firms instead of a few large ones. An instrument is available in the antitrust laws. It is not even necessary to pass new laws; it is necessary only to enforce the old ones.

This, for the technostructure, is a highly useful response. Those who detect fault can persuade themselves that they have a strong, even sanguinary solution. In extirpating monopoly they seem to go to the heart of the problem; in proposing the dissolution of the large corporation they seem to be unsparing as to remedy; in appealing to existing law they are being practical. And, some easily afforded legal harassment possibly apart, the technostructure can know that nothing will happen. The antitrust laws are now eighty years old; aside from some more recent legislation on mergers the basic structure of law is nearly sixty years old. Nothing has yet happened to arrest the development and burgeoning power of the technostructure.[6] A firm may, on occasion, be forbidden the acquisition of another firm; it may, on occasion, be required to divest itself of a subsidiary. But for more than half a century, if already large, it has been wholly secure in its existing size and almost wholly secure in further growth. Thus the remedy that emerges from the neoclassical model is harmless. It presents no threat to the power or autonomy of the technostructure or to its affirmative interest in growth. And since the remedy is thought to be comprehensive—since competition is considered the remedy for all industrial ills—it directs all complaint into an essentially harmless channel. What might be dangerous agitation for effective regulatory action or for public ownership or socialism comes out safely as a demand that the antitrust laws be enforced. And younger reformers can always be persuaded that the past fault lies not with the remedy but with the deficiency in courage, vigor—even perception—of the generation they are replacing. Only when they are too old to be troublesome do they discover that nothing has been changed. Best of all from the standpoint of the technostructure would be immunity from all attack. But the next best thing—and a very good thing—is a system of ideas that diverts all attack into channels that are safely futile.

[6] I am here summarizing an argument I have made elsewhere in more detail. See *The New Industrial State,* 2d ed., rev. (Boston: Houghton Mifflin, 1971), p. 184 et seq.

XIII

Costs, Contracts, Coordination
and the Tendency of Imperialism

THE PURPOSES OF THE TECHNOSTRUCTURE tell how it sets its prices. They govern also its behavior in the procurement of materials and equipment and of capital and labor. These decisions must reflect the protective purposes of the technostructure; the protective purposes being secured, the affirmative purposes are then served.

As the last chapter has shown, the major protective purpose of the technostructure where prices are concerned is served by having them under control. They cannot be market prices; these, by definition, are prices not under control. Control excludes price competition. It also allows general cost increases to be passed on to the consumer or buyer.

However the ability to pass on costs is only a protection for the technostructure of the particular corporation if the cost increase, in the manner of an industry-wide wage negotiation, affects all the firms of an industry at approximately the same time and by more or less the same amount. If the increase affects only one firm—if an oil company pays more for its crude or a steel company more for its ore while costs for the industry as a whole remain unaffected—it cannot count on being able to increase its prices. Other firms may not be cooperative. The protective purposes of the technostructure thus require that important costs also be under control. Equally important is the assurance of supplies at these controlled prices. The interruption or failure of the supply of some important material, component or skill is no less dangerous to the technostructure than an unplanned-for increase in its price.

2

Where the material or component is a major item of cost—crude petroleum, iron ore, pulpwood—the normal strategy of control is for the technostructure to reach back and possess its source of supply. This, at one move, brings both cost and supply under the authority of the firm. Everyday language emphasizes the security so provided. To describe a firm as a "fully integrated operation" is not to imply that it is more profitable but that it is safer—that it is more nearly invulnerable to accident, thus more secure in its earnings.

The need to secure raw material supplies has larger consequences. The imperial tendencies of the modern technostructure are extensively misunderstood by those who, using the oldest technique for economizing thought, substitute formula for fact. Economic imperialism is commonly associated with the need for markets. Marx saw capitalism as being dependent on a relentless expansion of sales and saw the colonial world as the primary servant of this purpose. Acceptance of Marx continues, as I have elsewhere noted, to be a litmus test for courageous, uninhibited thought.

The technostructure is, indeed, greatly concerned with overseas markets. But this concern is almost exclusively concentrated on markets in the other industrial countries. Partly this reflects the need for the international stabilization of markets, a matter also poorly understood and to be examined presently.[1] Equally or more important, economic development and rising living standards have made imperial concern for markets in the poor countries largely obsolete. In the last century the poor and colonial areas were worthwhile outlets for inexpensive tex-

[1] See Chapter XVII.

tiles, cheap metalware, railroad engineering and equipment and gimcracks, as Marx rightly emphasized. But these countries have, unfortunately, remained relatively poor. Markets there are still confined to a rather limited range of elementary consumers' goods and producer equipment. General Motors and IBM are deeply concerned with markets in Britain and Germany where the demand for automobiles and business machines is large; they have little interest in markets in the Union of Burma or the Republic of Chad where their market is effectively nil or in India or Indonesia where it is small. A line of argument holds that they pursue power (and shape foreign policy) out of capitalist perversity. They are, no doubt fortunately, not that indifferent to interest.

While economic development has tended to concentrate the concern for markets in the industrial countries, it has also increased the demand for raw materials. These are scattered over the planet more or less independently of economic development; accordingly the poor countries are of much greater interest as a source of materials. It follows that, so far as the technostructure has imperial tendencies in the poor countries, they are manifested much more on behalf of materials than of markets. Crude petroleum, iron ore, copper, bauxite, electricity for electrochemical purposes, natural gas and forest products are more important than sales of finished goods. To the extent that the corporation bends the foreign policy of the United States in the undeveloped countries to its needs, it is to its need for materials.

However here also there is danger of exaggeration. Consumption of raw materials has greatly expanded with economic development. In 1969, North America, with 6 percent of the world's population, consumed 37 percent of the world's liquid fuel, 37 percent of the world's total energy. The United States in that year consumed 24 percent of the steel, 41 percent of the rubber, 32 percent of the tin.[2] The quantities of materials

[2] *United Nations Statistical Yearbook,* 1970.

used in the last few decades vastly exceed the total consumption in all previous time. However supplies have shown a parallel if less publicized tendency to increase. Also technology has increased relentlessly the possibilities of substitution—synthetic for quarried nitrates, synthetic for natural rubber, aluminum for copper, one ferro alloy for another, plastics for everything. So, with the exception of oil, no very intense strategy has yet been required to get the requisite supplies. The tendency with materials is still, if anything, to abundance. Prices reflect this. And they are further kept cheap by the low labor costs and weak bargaining position of the supplying countries. There could be change but it is still ahead.

This defines the nature of imperialism in the Third World. It is an extension of the relationship between the planning and the market systems in the advanced country. As with the market system in the developed country, abundant supply, slight or no control over prices, a labor supply that lends itself to exploitation all mean intrinsically adverse terms of trade. The result is the same tendency to income inequality between developed and undeveloped countries as exists within the industrial country between the planning and the market systems. The planning system of the developed country is not chauvinist. It is indifferent in its exploitation of the domestic market system and the more general prototype of that system in the undeveloped country.

3

In the industrial country the need for a secure supply at a secure price is served, in the case of major materials and components, by integration—including possession of such services as pipelines for the transportation of oil or power for reducing alumina. It is also served by the contract. And this is far more important.

The contract can be thought of as extending the security which the large consumers' goods firm has in its own markets or the large weapons firm has in its relations with the government throughout the planning system, and to the common advantage of all concerned. The clearest case is that of a weapons manufacturer—the producer of a more arcane missile system or a new submarine. Rare accidents apart, its contract with the government gives it assurance on prices and amounts that will be sold. With this assurance it enters contracts with suppliers and they in turn with their suppliers extending down through many layers of subcontracts. These subcontracts assure the prime contractor on prices and supply. At the same time they give the subcontractor similar assurance on *his* prices and sales; they allow him to make commitments and otherwise undertake the planning requisite to fulfilling his contract. As earlier noted, the more technical the process and product, the longer the period between the original decision to produce and the emergence of the final product in remunerative quantity. And also the more technical the product, the more unlikely that the market can supply either the components, materials or labor that goes into its manufacture. With advancing technology, therefore, contracts increase in importance both for according protection over the longer period between initial decision and payoff and for ensuring the planning that will, in turn, ensure that needed materials, components and manpower are available when needed.

The case of the weapons firm is only the clearest example of how security is passed back into the system by contracts, and to the advantage of all concerned. The producer of consumers' goods—automobiles, television sets, appliances—does not have a contract specifying prices and amount to be sold. But it does have control over its prices. And, by the methods of which the next chapter tells, it does seek and obtain a substantial control over the behavior of its consumers. Having this, it then seeks to protect itself by contracts with its suppliers. And, having protection itself, it can afford the protection of a contract to its suppliers. This, then, allows them to invest and plan in order to meet the requirements of the original firm.

We have here, an important point, the essential mechanism for the coordination of production plans by different firms in the planning system.[3] The market, the traditional and revered mechanism for such coordination, does not work. As elsewhere noted, a higher price does not reliably accommodate supply to need within any predictable period of time, and this is especially so as products, components, materials and manpower become more specialized and technical. The contract, projecting the buyer's requirements for months and years—and specifying prices and terms—does ensure response. The firm's planning turns on prospective growth as the primary goal. From this, requirements are readily adduced. Along with similar information and guarantees from others the supplying firm is provided with the information required for *its* planning. It is then able to meet the needs of its customers in accordance with their schedule.

4

From the foregoing circumstances comes one of the most remarkable and also one of the most curiously unremarked features of the planning system. This is the vast web of interlocking contracts that it involves. The contract that secures the price and supply for one firm secures the price and sales for another firm. With increasing development and increasing technical complexity of products and the processes by which they are manufactured, this web of contracts continuously extends and thickens. In consequence millions of contracts are in existence at any time; tens of thousands are negotiated each week. Con-

[3] The absence of such coordination has recently been mentioned as a major oversight in my earlier argument along these lines. See Assar Lindbeck, *The Political Economy of the New Left: An Outsider's View* (New York: Harper & Row, 1971) and particularly Paul A. Samuelson's introduction thereto. In fact the mechanism of coordination along the lines above, and with particular emphasis on its importance as specialization and technology make the market ineffective, is discussed at length in *The New Industrial State,* 2d ed., rev. (Boston: Houghton Mifflin, 1971). Lacking, perhaps, was the uniquely explicit statement required for scholars whose instinct may, however innocently, be less to understand than to score points.

tract negotiation in the planning system is a major preoccupation that rivals concern for production or sales. A businessman, at any given time, is negotiating a contract, assembling the information that allows him to do so, contemplating the renewal of a contract or considering the cancellation of a contract. Business in the planning system, it can be said with only slight exaggeration, is mostly contract negotiation.

The contract accorded by the manufacturer or supplier of end products to those that supply him solves, or goes far to solve, the problem of vertical coordination in the planning system. The firm producing automobiles or weapons uses its power to plan its own output; by use of contracts it allows those on whom it depends to plan their production and thus ensure that the requisite components are forthcoming when and in the amounts needed. There remains the problem of coordination as between end products. The sale and consumption of numerous of these, as in the case of electricity and air conditioners, is interdependent.

In the neoclassical system failures of coordination as between industries cannot occur. Should there be such a tendency, price will accommodate demand to supply or vice versa. In the fully planned socialist economies such failures are commonplace—they are, indeed, one of their trademarks. That these failures do not occur in the planning system, as here described, has been taken by reluctant scholars as a reassuring indication that the system does not exist.[4]

It is not a happy defense. Circumstances as usual have been exceedingly uncouth. The modern planning system is, in fact, plagued by such problems of coordination. And the layman or unconditioned student will not resist the thought that they are becoming more serious. The electrical utility industry, pursuing its more difficult goal of growth, is no longer able to keep pace with the more dynamic electrical products industry. The consequence is brownouts and blackouts. The expansion

[4] See Samuelson introduction to Lindbeck, *Political Economy*.

of the oil industry, although powerful, no longer keeps pace with the growth of the automobile industry or the output of other petroleum-using products. Practical manifestations of this may well be evident by the time this book appears. The supply of railroad rolling stock, provided by a weak industry, no longer matches general need. These problems in coordination, given the nature of the planning system and the uncoordinated thrust for growth in different industries, will recur and multiply. They are inherent in the system. They are another potential source of disturbance to the neoclassical comfort.

5

The contract is central for the protection of prices and costs and of sales and supplies at these prices and costs. This security having been provided, the affirmative purpose of the techno-structure becomes relevant. This will be, for each firm in the planning system that is a party to the contracting process, a level of price or cost which best serves growth as modified by the need to show improvement in earnings.

This purpose, in turn, is what makes the great web of contracts practical—as it would not be if the neoclassical system mirrored the reality. In the neoclassical system prices of final products are those that maximize profits. Of the resulting profit everyone seeks as much as possible. This means that negotiation between a producer and his suppliers is, essentially, to divide a pie. What one gets the other loses. Were this the case, such negotiation, a zero sum game, would be a time-consuming test of relative power, endurance and cupidity. Under such circumstances there would be no hope of accomplishing the myriad of transactions presently completed. Most would end with a dissatisfied participant. Instead of the friendly glass that now concludes a bargain, the more normal outcome would be aggravated assault by the drunk who lost.

These difficulties and unpleasantries do not arise. That is because, once the protective purposes of the participants to the

negotiation are served, negotiations have ultimately to do with
establishing the level of cost or price which maximizes growth
for both participants. For buyers and sellers of roughly similar
power—parties to the bargain that are more or less equally in
need of each other—these prices and costs tend to be the same.
Subject to minimum income constraints the merchant ore firm
wishes to maximize deliveries of ore, and the steel company
wishes to maximize sales of wide sheet for automobiles. Both
seek a price for the ore consistent with this goal; it is, roughly
speaking, the same price for both. The steel company, again
seeking to expand sales, then negotiates similarly with an automobile company that is concerned to maximize sales of cars.
Again there is a common interest. Negotiation, where growth
is the goal, is not a zero sum game; the increase in sales, when
the price is right, is something in which all have a share. For
this reason contract negotiation in the planning system is not
unduly difficult. It is accomplished between friendly men who
are concerned, primarily, with reconciling differing assess-
ments of the same interest. Again the merit of a valid view
of economic processes is that it explains the inexplicable.

6

Where there is a difference in the power of the negotiants, the
outcome is not more difficult. But it is different. And the difference explains one of the most fundamental tendencies in income distribution among different parts of the economic
system.

Size, as usual, bears heavily on the case. The large firm can
turn to alternative sources of supply. The case of the large
weapons firm apart, it is rarely dependent on one customer.
The smaller firm has fewer options. On frequent occasions—
the appliance manufacturer who serves Sears, Roebuck is an
example—it is limited to one customer. When a firm that has
alternatives does business with one that does not, the relationship, as also the resulting bargain, will not be equal.

The nature of the inequality needs, however, to be understood precisely. The smaller firm will invest more adequately and thus produce more economically if it has the security of a contract on which it can reliably survive. The larger firm derives no advantage from negotiating a price lower than that at which the smaller firm can continue to supply the product. A contract that is so unfavorable or so inflexible that it destroys the small firm is self-defeating. The effect of power emerges in the way price is graded to need. The larger firm can calculate the income that the smaller firm requires for survival and minimal satisfaction of its affirmative purposes, and it does so as a matter of course. The small firm can make and enforce no similar calculation on the larger firm. The consequence is that a smaller firm doing business with a larger one will almost always have its returns more nearly at the necessary minimum than the larger firm doing business with smaller ones.

The presumption, thus, is to inequality of return as between large firms and small.[5] The foregoing applies to large and small firms within the planning system. The presumption of inequality of return is greatly enhanced if the small firm does not have control of its prices or costs. And it is similarly increased if there are circumstances which cause entrepreneurs or workers to lower their rate of return, more or less without limit, in order to remain in business—if there is self-exploitation. We have seen that there are numerous such circumstances. In part they are what differentiate the planning system from the market system.

The market system is a world of small firms. It also sells a great many things—agricultural products, forest products, other raw materials, numerous services and many parts and components—to the firms of the planning system. Where the industry approaches the classically competitive structure, contracts are not commonly used. Supply responds reliably to

[5] This, it must be emphasized, *cannot* be measured by comparing profits. Salary earnings reward the technostructure, and in the large firm the goals of the technostructure take precedence over those of the owners. Inequality between firms is thus manifested not alone in differences in profits but also, and even more, in the difference in ability to pay good salaries.

changes in market prices, and, sellers being numerous, it is neither practical nor necessary to seek arrangements with each one.[6]

The market system, as we have seen, allows of self-exploitation and encourages it as convenient social virtue. And not only is independent entrepreneurship held to be, at least in part, its own reward; rather more improbably it is held to be a substitute for the protection of unions and minimum wage legislation and thus of income for those, like farm workers, who are closely associated with it. Technology being simple and capital requirements modest, entry into the business is easy. So the man who cannot find employment in the planning system can often become an entrepreneur in the market system.

Finally, agriculture is in the market system. The birthrate is substantially higher in the rural than in the urban states— 20.0 per thousand in 1969 in Mississippi as compared with 16.4 in Connecticut.[7] Additionally, in consequence of the socialization of agricultural technology and the rapid rate of technological innovation in the supporting industries, agricultural productivity—output per worker—has increased more rapidly in recent years than productivity in urban pursuits. It follows that if agricultural and industrial output are now increasing at more or less the same rate, there must be a steady migration of workers from agriculture to industry. Otherwise agriculture will have unemployed or otherwise redundant workers. For many the alternative to migration is to remain and accept a lower wage.[8] All of these factors serve to depress income in the market system as compared with the planning system.

[6] However, in poultry production, cattle feedlot operations, hog production, there is a growing tendency to give the small entrepreneur the security of a contract. Investment that would not otherwise be feasible thus becomes possible.

[7] *Statistical Abstract of the United States,* 1971 U.S. Department of Commerce.

[8] In recent years (1964–1968) the migration from farms was sufficient to raise the median income of the families that remained in relation to all families. But the increase (from 0.54 to 0.67 percent of income of all families) left those in agriculture still far behind. See Andrew F. Brimmer, "Inflation and Income Distribution in the United States," *The Review of Economics and Statistics,*

What was previously a provisional conclusion will now be fully evident. In the market system managers and workers continue to supply products and services at levels of remuneration that are below those for comparable talent in the planning system. And this is a durable condition. It follows that equality is not the tendency between the planning and the market systems; the basic tendency is to inequality. The figures affirm the expectation. In 1971, hourly compensation in durable-goods manufacture, the part of manufacturing most strongly represented in the industrial system, averaged $3.80. In nondurable manufacturing, where (in apparel and other manufacture) there is a substantial market component, it was $3.26. In services, which are strongly oriented to the market system, it was $2.99. In retail trade, where the market system also retains a strong foothold, it was $2.57. In agriculture, the industry most characteristic of the market system, it was $1.48.[9] Were executive and entrepreneurial income included with that of wage labor, the difference would, of course, be greatly increased.[10,11]

The relationship between the planning and the market sys-

Vol. 53, No. 1 (February 1971), pp. 37–48. As later noted, acute inflation could work at least briefly in favor of market incomes.

[9] Economic Report of the President, 1972. (Preliminary figures.)

[10] Peter Henle's important study, "Exploring the Distribution of Earned Income," U.S. Department of Labor, *Monthly Labor Review* (December 1972), p. 16, shows increasing inequality for both "individual occupational and industrial groups" and concludes that "changing occupational structure, itself the product of technological advances, has contributed to inequality generally, and more specifically in those industries experiencing rapid increases in the number of highly paid professional and managerial personnel." Needless to say, the latter industries are in, and in large measure comprise, the planning system.

[11] In industry generally the large plants characteristic of the planning system regularly pay higher wages than the small plants characteristic of the market system. See Stanley H. Masters, "An Interindustry Analysis of Wages and Plant Size," *The Review of Economics and Statistics*, Vol. 51, No. 3 (August 1969), pp. 341 ff. T. Paul Schultz has recently concluded that between 1939 and 1969 inequality among fully employed workers increased among men and women less than 25 years old but was otherwise relatively constant. "Long Term Changes in Personal Income Distribution: Theoretical Approaches, Evidence and Explanations," *The American Economic Review*, Papers and Proceedings, Vol. 62, No. 2 (May 1972), p. 361.

tems, their unequal rate of development, the exploitation of the second by the first, the resulting inequality in return are central features of the modern economy. They are, accordingly, a central concern of this book.

Neoclassical doctrine does not challenge the validity of the figures on hourly earnings just mentioned. It holds that resources move from lower to better paid employments in an economy homogeneously composed of broadly similar firms. This reduces inequality. This minimizes or eliminates the need for corrective action; one is not moved to remedy what time and the nature of the system itself will cure. Indeed the neoclassical learning rebukes farmers in the market system for seeking government support for more effective bargaining with the system. And it rebukes retailers and other small businessmen for organizing to defend margins or fix prices. Such entrepreneurs are not seeking to redress their inherent weakness, as we here see; they are trying to create monopoly. The effect of these beliefs is to disguise and so to perpetuate the income advantage of the planning system. The neoclassical model, in its education, is again performing an instrumental function.

XIV

Persuasion—and Power

Mr. Hill wanted more women to smoke Lucky Strikes: research showed that sales to them were down because the green-packaged cigarettes clashed with their costumes. "Change the color of the package," I suggested. Mr. Hill was outraged. I then suggested we try to make green the dominant color of women's fashions . . . For a year we worked . . . Green became fashion's color.

—Edward L. Bernays
The Business History Review
Autumn 1971

WE COME NOW to a decisive point in the development of a modern view of the economic system. The neoclassical model concedes that producers in many industries have a substantial measure of control over prices and costs. Such is the nature of monopoly and oligopoly. So long as this is to maximize profits, and is not part of any more comprehensive exercise of power, the firm remains subordinate to the ultimate will of the user of the goods. As the tastes and needs of the user change, so does the amount he will take and the price he will pay. In responding to these changes, as it must if it is to keep its profits at a maximum, the firm responds to the user's authority. Although the response is imperfect, the sovereignty of the user remains unimpaired.

This admirable vision of the ultimate power of the user cannot be sustained, however, if his tastes and needs fall under the authority of the producer. This does not require much explanation. The consumer is not sovereign if he or she is subordinate, or partly subordinate, to the will of the producer.

That the economy is ultimately in the service of the consumer cannot be believed if the producer can manage the consumer —can bend him to his needs. And once it is agreed that the producer has a measure of authority over the consumer or other user of his goods, the way is open for a further and massive breach in the dike. For then it can be argued that the control of prices, costs, consumer demand and the state is all part of a single deployment of power—one that serves the purposes of the technostructure in particular and the planning system in general. Exponents of the neoclassical system unite, not without a certain scholarly fervor, in denying that the producer has effective power over the users of his products. Once again their instinct, viewed in the light not of truth but of self-preservation, does not serve them badly.

Yet no myth, however serviceable to particular interest, is wholly satisfactory if it taxes belief. In the exercise of monopoly power there is control of prices and, where possible, of costs. The existence of this power is accepted in the traditional or neoclassical view. But the profit maximization of the monopoly would be most negligently served, as almost all must agree, if, the initial power over prices having been secured, no effort were made to affect the demand for the product—if the firm then relaxed and contented itself with the whimsical acceptance or rejection of its product by the consumer. The obvious counterpart of the control of prices is an effort to control the response of buyers to those prices. A strategy for protecting established belief which seeks to confine corporate power to the control of prices and costs, however great its service to intellectual vested interest, is also absurd.

2

The final users of goods and services are private consumers and the government. Efforts to influence demand extend comprehensively to both. The management of the private consumer has two dimensions: There is the preference or nonpreference of the consumer for the producer's product or service. This must be favorably influenced. And there is the equally poignant question, given that preference, of whether the consumer has the income to buy the product or service. Not much is gained by persuading customers to a product if they cannot afford it. An effective strategy for ensuring the desired response of the private consumer must be concerned therefore both with influencing his attitude toward the particular product or service and ensuring, insofar as may be possible, that he—and consumers generally—have the wherewithal or effective demand to buy the product.

The management of the private consumer of goods is inextricably associated with the management of the public demand for goods. The corporation seeks to manage the choices of the private consumer. It seeks also to manage the purchases of the state. The techniques are radically different in the two cases; the purpose is the same. And there are further interrelationships. The public expenditures that are the product of the influence of the technostructure on public procurement are also important for sustaining the flow of public expenditure that stabilizes the purchasing power of the private consumer. Military expenditures in particular both buy the products of the supplying firms and sustain demand in the economy as a whole. It is thus seen that what economists compartmentalize as macroeconomics and microeconomics are parts of a larger whole, one that is formed by the power of the planning system.

The management of demand requires management of the state for yet other reasons. Some kinds of private demand are only possible if there is complementary action by the state—the demand for automobiles requires facilitating public expenditure for highways; demand for air travel and aircraft requires public expenditure for terminals and airways. And broad patterns of consumption are established by public policy. People go to work by automobile in the United States partly, no doubt, because of preference but partly because no alternatives exist.

The use of public resources for alternative modes of travel has been powerfully discouraged by automotive interests.[1]

Although we are here concerned with tightly interrelated phenomena, it will be convenient to look first at the way in which the planning system brings its power to bear on the private consumer. Then we shall see how it influences public purchases of its products and otherwise arranges needed public action. The problem of stabilizing demand in general, though a part of the same process, will be postponed to a later chapter, where its bearing on the market system can also be seen.

3

The management of the private consumer is a task of no slight sophistication; the cost is considerable, and it uses some of the most expert and specialized talent to be found anywhere in the planning system. Its most obvious instrument is advertising. And the uniquely powerful instrument of advertising is television which allows of persuasive communication with virtually every user of goods and services and with no minimum requirement in effort, literacy or intelligence. But the management also involves the deployment of sales and merchandising staffs and of sales and dealer organizations. It makes extensive use of market research and testing to ascertain to what the consumer can be persuaded and by what means and at what cost. It extends deeply into the choice and design of goods to ensure that they incorporate features that lend themselves to persuasion—that have good selling points. It makes extensive use of innovation which, as we shall presently see, differs sharply from the classical purposes of invention, which were to serve some need the inventor had perceived and sought to fill. Modern innovation is more often to create a need that no one had pre-

[1] The extent to which consumer choice is affected by such denial of alternatives has been brought to my attention by Paul Sweezy, who has taxed me, I think rightly, for neglecting it in earlier work. See his "Comment," *The Quarterly Journal of Economics,* Vol. 86, No. 4 (November 1972), p. 661 et seq.

viously perceived. Or it exploits the close association that exists in the public mind between innovation and improvement. To the peculiarities of modern technical innovation we return in the next chapter.

A special word is required on market research. This, it has occasionally been argued, is to ascertain what consumers want. In consequence its existence affirms the ultimate power of the consumer and ensures that production is more efficiently subordinated to that power. As often or more often it is to ascertain the effectiveness of different kinds of persuasion or how well different products, brands or packages lend themselves to such persuasion. From this the firm learns how money for persuasion can be most efficiently spent—what sales effort gets the best results and what products lend themselves best to persuasion and how much. Such effort hardly affirms the sovereignty of the consumer.

Much of what is called market research, it may also be noted, is imprecise. Subjective, random or fraudulent judgments are offered in impressive pseudosociometric tables to suggest a precise relationship between outlay on different kinds of persuasion and the resulting sales. This will not be surprising. An industry that employs much carefully tempered mendacity will not be sensitive to its application to itself.

The power to influence the individual consumer is not, of course, plenary. It operates within limits of cost. The winning of customers or custom will generally be at increasing cost; the shape of this function (the curve showing the cost of obtaining each added increment of sales) will depend on the nature of the product and the quality of the persuasion, and it will be subject to sharp variation over time. Both the position and stability of this function will depend on past outlays for persuasion. What will have to be spent to win a given amount of custom for a particular soap will depend on what has been spent on that soap in the past and also on what has been spent by all soap manufacturers to establish the imperatives of a clean and odorless personality.

That the power to manage the individual consumer is imperfect must be emphasized. Scholars who, for traditional and instrumental reasons, argue that the sovereignty of the consumer is unimpaired normally hold that the alternative to total independence of consumer choice is total subordination of that choice to the will of the producer. The consumer, not being wholly sovereign, is wholly a puppet of the producer. It is necessary only to outline this dialectic to indicate its tendentious design.

4

The protective purpose of going beyond prices to influence consumer response is to prevent the defection of consumers which thus would plunge the firm into loss. The affirmative purpose is, of course, to recruit new customers and thus to expand sales —to serve the goal of growth. As sufficiently remarked, the typical industry of the planning system consists of a few large firms. This means that sales can be expanded both by recruiting new users and by persuading customers of other firms to switch. Other firms, for their part, will be engaged in the same effort. The management of the consumer is thus an intricate complex of efforts to recruit new users, win the customers of other firms and hold existing customers in face of the corresponding efforts of the other firms. Since the gain of customers by one firm means their loss by another, the affirmative and protective purposes of consumer management, as they are actually pursued in any consumer industry, are in some degree in conflict.

This, however, is only partially so. For firms that have the scale and resources to participate fully in the persuasion[2]—in

[2] See William S. Comanor and Thomas A. Wilson, "Advertising and the Advantages of Size," *The American Economic Review*, Papers and Proceedings, Vol. 59, No. 2 (May 1969), p. 87 et seq. The authors conclude that the large firm enjoys significant advantages in the use of advertising. In an earlier article ("Advertising Market Structure and Performance," *The Review of Economics and Statistics,* Vol. 419, No. 4 [November 1967], pp. 423 ff.) the same authors also found a general association between advertising expenditure and profits.

the automobile industry, General Motors, Ford and perhaps Chrysler as distinct from American Motors or such earlier casualties as Studebaker and Packard; in the soap and detergents industry, Procter & Gamble, Lever Brothers, Colgate-Palmolive—the aggressive and defensive operations of the participating firms come eventually into a rough equilibrium. The company that is gaining rests with its existing system of persuasion; the company that is doing less well seeks more effective means of persuasion as to existing products or searches for products or designs that lend themselves more effectively to persuasion. Sooner or later it succeeds, and this returns the play to its previously more successful rivals. The result is a control of consumer reactions which, though imperfect and greatly complicated by the rivalry, is still far more secure than would be the ungoverned responses of consumers in the absence of such effort.

Meanwhile the aggregate result of the effort is solidly beneficial to all the participants in the planning system. It recruits new customers for all firms in the particular industry, associates existing customers more firmly with the products of the industry and powerfully advances the larger purposes and values of the planning system. These are vital services and deserve a special word of summary.

The advertising of the individual automobile company seeks to win consumers from other makes. But the advertising of all together contributes to the conviction that happiness is associated with automobile ownership. Additionally, make and model apart, it persuades people that the contemporary tendencies in automobile physiognomy and decoration are desirable, that those of the past are obsolete, eccentric or otherwise

---

Roughly the same conclusion is reached by Leonard W. Weiss in "Advertising, Profits and Corporate Taxes," *The Review of Economics and Statistics,* Vol. 51, No. 4 (November 1969), pp. 421 ff. There are various possible explanations for this; the most obvious is that aggressive demand management serves effectively to protect and enhance returns. The notion of an equilibrium as here outlined follows the argument advanced in *The New Industrial State,* 2d ed., rev. (Boston: Houghton Mifflin, 1971), p. 204 et seq.

unworthy. Thus it encourages the general discarding of old vehicles and the purchase of new. Similarly, if one soap manufacturer can establish that white sheets are an index of womanly
virtue, this virtue is rewarding to all soap and detergent manufacturers. If one manufacturer can make modest intoxication
a mark of suave respectability, so it becomes for all makers
of intoxicants. If one hairdressing contributes to successful se-
duction, then so may all.

More important still, the aggregate of all such persuasion
affirms in the most powerful possible manner that happiness
is the result of the possession and use of goods and that, pro
tanto, happiness will be enhanced in proportion as more goods
are produced and consumed. Thus the persuasion proclaims
and extends the values of the planning system in general and
its commitment to growth in particular. This helps also to support its claim for assistance from the state on behalf of its needs.
One branch of neoclassical economics has long held that the
advertising and persuasion in the typically oligopolistic industry is a purely wasteful exercise in aggression and defense—
"a form of nonprice competition . . . of a mutually neutraliz-
ing, standoff sort, with no technical or social benefits at all."[3]
The only consequence is higher prices to the public or lower
earnings for the participants.

Were this the case, steps would long ago have been taken
to limit advertising outlays by common agreement. No law
would have stood against this effort, for the cost to the industry and the waste for the public would have been solemnly
and influentially cited and policy would thus have been accommodated to the needs of the planning system. In fact, competitive persuasion serves the common purposes of the planning
system. Accordingly no important effort has ever been made
to limit it.

[3] William G. Shepherd, *Market Power and Economic Welfare* (New York:
Random House, 1970), p. 53. I do not imply that Professor Shepherd, to
whose competent work all who are concerned with these matters are indebted,
is by any means a captive of the stereotypes.

5

The technostructure also extensively influences the purchase of public goods in accordance with its needs. Here, however, there is general recognition that the orthodox economic view, though ceremonially presented to the young as a conventional characteristic of democracy, has little relation to the reality.

In the orthodox or traditional view the choice between private and public goods and services and between different public goods and services is expressed, indirectly, in the choice of candidates and party for public office—in the choice between those avowing an interest in more or less taxation and, given the level of public expenditure, between those urging greater or less emphasis on education, welfare, public works or other public goods and services. Among the latter services are weapons and weapons development, but they are singled out for no special attention.[4] The candidates so chosen relate the public will to the executive through their power over legislative authorization and appropriations. The executive—the bureaucracy—is the passive servant of the legislature and thus, ultimately, of the citizen. In the United States the election of the Chief Executive, who exposes his position on these matters to the voters, further reinforces the citizen control.

Few who teach this doctrine would be willing to admit personal belief; to do so would be to damage one's reputation for a minimally reputable skepticism. In the case of weapons—missile systems and missile defenses, nuclear aircraft carriers, fighter aircraft, manned bombers—the process, it would be agreed, almost exactly reverses the orthodox formula. The ini-

[4] "Introductory courses in economics, as reflected in principal textbooks used in American colleges and universities, usually do not recognize the existence of the military-industrial firm or a war economy. In these texts the magnitude and the characteristics of military economic activity in the United States since the Second World War either are not mentioned at all, or are dealt with in a few sentences or paragraphs." Seymour Melman, "The Peaceful World of Economics I," *The Journal of Economic Issues*, Vol. 6, No. 1 (March 1972), p. 1.

tiating decision is taken by the weapons firm and by the particular service for which the item is intended. The action is ratified by the President who, though not without power, is extensively a captive of the bureaucracy he heads. The Armed Services Committees of the Congress, staffed with reliable sycophants of the weapons firms and the services, accept all but automatically the decision so taken. The role of the rest of the Congress is minimal; that of the public is nil.

The foregoing as distinct from the doctrinal view accords with what would be expected from the present analysis, given our knowledge of where power is located and how it is used. Some would assign a special primacy to the weapons firm. This derives from a view of capitalism which automatically accords a commanding role to the capitalist firm. However two sets of bureaucracies, the technostructures of the weapons firms and that of the Pentagon, are involved. It cannot be assumed that the one kind of organization is less powerful than the other.

Rather they jointly pursue common interests in growth and technical innovation—they have the same relationship as that between private technostructures described in the last chapter. Members of both the public and private bureaucracies are served by growth and the consequent promotions, pay, perquisites, prestige and power, and what expands one bureaucracy expands the other. Technical development, as the next chapter will show, is particularly important both for the autonomy and growth of the public bureaucracy and for the supplying technostructure. Here, accordingly, the reciprocal support is especially great. The service, often with the help of the weapons firm, defines the need for the product; the firm then undertakes the development. Both gain.

There will be a similar tendency to reciprocal support whenever a technostructure and a public bureaucracy are closely juxtaposed. Such is the relation between the Atomic Energy Commission and its supplying industries. Such is the relation, as regards roads, between the Department of Transportation and the automobile industry. Even where there is a presumptively adversary relationship between a public and a private bureaucracy, as between the Federal Communications Commission and the television and broadcasting networks, reciprocal support is possible.[5] This tendency for the public and private organizations to find and pursue a common purpose is of sufficient importance to justify a name. It may be called Bureaucratic Symbiosis.

In the United States bureaucratic symbiosis reaches its highest state of development in the relation between the weapons firms and the Department of Defense and its constituent elements. Lockheed, Boeing, Grumman or General Dynamics can develop and build military aircraft. This serves their affirmative goal of growth with the concurrent reward to their technostructures. The public bureaucracy that is associated with research and development, contracting, contract supervision, operations and command is similarly rewarded by the development and possession of a new generation of planes. But bureaucratic symbiosis also works at a more elementary level. The technostructure of the weapons firm is a natural source of employment for those who have completed a career in (or otherwise exhausted the possibilities of) the public bureaucracy. Leadership in the Department of Defense, by the same token, is extensively in the hands of men recruited temporarily from senior positions in the technostructures of the weapons firms. Not only is this exchange rewarding to individuals, but it serves, more than incidentally, to cement the symbiotic relationship.

In the symbiotic relationship between the public and private bureaucracies, it may be stressed once more, no conclusion can or should be reached as to where the initiative lies. Certainly no one can say with assurance that it is with either the public bureaucracy or the firm. What is clear is that the initiative does not come from the citizen. More completely even than in the case of the consumer, effective power has passed to the producers—either to the producer of the weapons or the producer

[5] See Chapter XVI.

of the military service that employs the weapons. And, as noted, even pretense to the contrary is no longer quite respectable.

6

Obviously the power of different producers in relation to the consumer or citizen varies greatly. It is greatest where the development is most advanced—where the firm is largest and its technostructure is most fully developed. In the private sector of the economy it is greater for the automobile, soap, tobacco and manufactured-food industries than in, say, housing, medical care or the arts. In the market system the power of the producer becomes minimal or disappears. In the public sector the power of the producer will be greatest, that of the citizen least, where there is bureaucratic symbiosis—where a large aerospace firm works symbiotically with the Air Force. It will be least where small construction firms build low-cost housing for a local housing authority or funds are supplied to a local school district.

Thus the conclusion, already suggested and now becoming firm. How economic resources—capital, manpower, materials—are allocated to production, both in the private and public sectors of the economy, depends, and heavily though of course not exclusively, on producer power. And, with the development of the economy, it depends increasingly on that power. This is a basic tendency of the economic system.

In the neoclassical model production is controlled by consumer and citizen choice. The ultimate equilibrium corresponds to their need as interpreted by themselves and made effective by their income. In the modern reality the equilibrium reflects the power of the producer. This, not "need" in any exclusive or dominant sense, controls what the economy does. Production is great not necessarily where there is great need; it may be great where there is great capacity for managing the behavior of the individual consumer or for sharing symbiotically in the control of the procurement of public goods and services,

all in the interest of bureaucratic growth. This is in sharp contrast with the neoclassical view of power which holds that
power restricts output in the manner of the classical monopoly.
But a moment's thought directed to the areas of abundance
in the economy—automobiles, weapons, soaps, deodorants and
detergents—will suggest that the present analysis is not in conflict with common observation and common sense.

Discussion of the practical issues associated with this over-
production—and the underproduction of other things—can only
be avoided if it is agreed that the consumer and citizen are
resistant to the power of the producer; if advertising and salesmanship are the froth and not the substance of economics; if,
as an expression of oligopolistic rivalry, they cancel out; and
if the great weapons firms, though admittedly powerful, are
a flaw, sui generis, left over from the Cold War. If economics
helps with such belief, it is, from the viewpoint of those who
exercise power, a most benign thing. If it insists on identifying
the exercise of power that explains the reality, it is less benign.
Questions as to the legitimacy of power follow. So do questions
as to the effects of its use. The need for remedial action to
align the use of power with the public interest can no longer
be escaped. And such remedial action ceases to be exceptional
but becomes, instead, an intrinsic need.

XV

The Revised Economics of Technical Innovation

IT IS NOW POSSIBLE to see with some clarity the role of technical innovation in the modern economy and in the planning system. This is a matter of considerable interest. Few things have been more striking than the revision in recent times in public attitudes toward technical change. Until very recent years it was an absolute social good. Only eccentrics questioned it. The word invention was synonymous with progress. Engineers and scientists, the sources of such innovation, were prime social benefactors. The promotion of scientific and technical innovation was a valued and unquestioned function of the state.

Now doubt is commonplace. Much innovation in consumers' goods is felt to be fraudulent. It is taken for granted that many much-heralded inventions will have as their most striking feature that they do not work or they will prove hazardous. The social movement called consumerism and dramatized by Ralph Nader owes its origins in no small part to these characteristics of innovation. And increasingly it is believed of innovation that, though it serves its function—though it moves people at supersonic speeds or destroys incoming missiles—it is impressively negligent of the consequent social damage or public danger. Given a sufficient rate of technical progress, it is increasingly assumed, all the beneficiaries will be dead.

In the neoclassical model technical innovation was of two

kinds. It brought into existence new or improved products or services that were welcomed and bought by consumers because they better served their needs. Or such innovation improved the processes by which products were made or services rendered. (In technical terms innovation operated either to create or alter demand functions or to lower cost functions.) The invention or improvement of products was always in response to a *perceived* need of the consumer. Emerson's advanced mousetrap to the contrary, the consumer might need to be informed of the improvement; perhaps some persuasion would be necessary to overcome his innate conservatism. But the invention had merit because it identified a need. The persuasion revealed the relation to need; it did not create the need. Invention or improvement of processes reduced cost and ultimately also prices. Of the advantage of this no one could be in doubt.

Since innovation provided the individual with better or cheaper goods, the neoclassical model, not surprisingly, held it in the highest regard, and the disapproval of any interference therewith was stern. Workers might resist new processes because they feared loss of their jobs. Producers might seek the suppression of both products and processes because they feared the obsolescence of their investment. In both instances the public interest in cheaper or better products suffered. This being so, such interference—all opposition to "technological progress" —was held to be uniquely misguided. So, generally speaking, it is still regarded.

In the planning system, innovation, like other activity, is highly organized. The thing to be invented or the improvement in process that is to be made is commonly settled on in advance. The development, with rare exceptions, is pursued in accordance with established timetables and within approved budgets. The notion of a totally spontaneous invention by an individual deriving from a brilliant, innovating thought is not quite dead. It survives in part because, being inexpensive and independent of organization, such invention is possible for the small firm and thus for the market system. Without it innovation of all kinds would, in theory, be the exclusive property of the planning system with its resources in specialized knowledge, organization and capital.

Most innovation does, in fact, require such specialized knowledge, organization and financial support. On this—that the great bulk of expenditure for research and development is by larger firms—there is no disagreement. What remains to be seen is that as innovation becomes organized and deliberate, it also comes fully into the service of the technostructure. And because it now serves the technostructure, not the consumer, it will, as a quite normal consequence, be in conflict with public purpose.

2

Innovation, as it is brought to bear in improving manufacturing or other processes as distinct from innovation in things made or services rendered, serves two purposes of the technostructure. It reduces costs and thus allows prices that encourage greater sales. In this fashion it serves the affirmative interest of the technostructure in growth.

Innovation in processes also enhances the power and security of the technostructure and thus serves its protective purposes. This service is a trifle more complex.

In the modern corporation, as previously noted, the factor of production that is not fully under the control of the technostructure is the labor force. This remains a possible challenge to its authority. This challenge, normally that of a union, is extensively neutralized by the convention that excludes union interference with what are called the prerogatives of management. It is further neutralized by the firm's control of its prices, by industry-wide collective bargaining and by the tacit understanding among firms that wage increases will be passed on to the public. Neither the industry nor any single firm is thus threatened by wage increases that it cannot pass on.[1]

Innovation in productive processes all but invariably involves the substitution of capital for labor. In the planning system the savings that are the source of capital come extensively from the earnings of the firm—its supply of capital is at its discretion and under its control. Prices of the resulting machinery and equipment are more predictable than wage costs, and, once installed, machines do not go on strike. Both the cost and performance of capital are thus far more reliable than those of labor. So innovation and the concurrent substitution of capital for labor increase the security of return of the firm and so serve the protective purposes of the technostructure. This means, in practical terms, that the question of labor-saving machinery in the modern corporation is not uniquely a matter of pecuniary calculation. It is quite conceivable that the substitution of capital for labor will occur at increased cost. This is for the wholly rational reason that the substitution of capital for labor accords enhanced security and power to the technostructure. It allows of more complete planning by the planning system.

From the foregoing characteristics of process innovation come two consequences. The first is for the labor force in the planning system to shrink in relation to output as compared with the market system. And this shrinkage may be greater than that resulting from the economies of the innovation. This could mean that the common defense of technical innovation and the common justification of the associated labor displacement are in part a fraud. The common justification holds that the displacement is always in the interest of lower costs—that

[1] We may note that a consequence of innovation in processes is that the resulting cost reduction reduces the amount of the wage increases the planning system must pass on in prices. This, in turn, favors its affirmative purpose of growth. In the market system, agriculture apart, productivity gains are generally lacking. That is why wage increases in the service industries that parallel those in the planning system will eventually occasion much larger increases in prices.

the discomfort of the workers so displaced is offset by the gains in cheaper products. The real reason could be not lower costs but enhanced security and power for the technostructure. This is a more difficult matter to defend; even a more than normally acquiescent union leadership might have difficulty accepting technical innovation and resulting labor displacement if the purpose is the replacement of obstreperous humans by more costly but more compliant machines.

The second consequence of such process innovation is its effect on the environment. Some technical innovation is damaging to surroundings. It cannot be assumed, however, that new technical processes are always more adverse to the environment than old ones—that the thermal pollution from an atomic energy plant is worse than the smoke from an old coal-fueled plant or that an airplane passing noisily overhead is worse than an express train roaring through houses huddled (as often they were) a few feet from the tracks. But any new form of environmental damage, because it is new, will always seem worse than the one to which the community has become accustomed. The noise of jet aircraft, being new and having a new clientele, will seem worse than that of the trains. Thermal pollution, being more mysterious, will seem more insidious than sulfur and soot.

Technical innovation in processes serves, we have just seen, the affirmative goal of growth. That growth adds to pollution of air and water and to other environmental disharmony. Since process innovation will often involve construction of a new plant or use of a new location, it is regularly singled out for blame that is properly attributable to the technostructure's pursuit of the goal of growth itself. In later consideration of remedial action this is a matter of major importance. It is the pursuit by the technostructure of its own goals, exercising its own power to do so, and not technological innovation per se that is at the heart of the environmental problem. We turn now to the role of product innovation.

3

As product innovation becomes organized and passes under control of the technostructure, it too is made to serve the purposes of the technostructure. The primary affirmative purpose being growth, the primary question that is asked about innovation is whether it will add to sales. To serve this purpose it no longer need serve a previously perceived need of the consumer; it is necessary only that the innovation contribute to the larger process by which the consumer is persuaded. Usefulness, once indispensable to the success of an invention, now becomes only one of several requisites of such success.

Newness quite apart from any function may be highly serviceable to persuasion. The popular view of invention has long been strongly linear—there is a powerful presumption that a newly invented product is better than something that was invented a year or ten years ago. The last thing is the best. This belief, in turn, is derived from the genuine experience of the past. When inventions succeeded or failed to succeed in accordance with whether they met or failed to meet needs perceived by the user, later inventions were better than earlier ones. Inventions which were not better sank without trace. People, not surprisingly, continue to associate newness with improvement. Economic and other pedagogy reinforces the point. In all instruction, invention continues to be synonymous with usefulness. Next to his experiments with the kite and lightning Benjamin Franklin is most celebrated as the founder of the Patent Office. Leonardo is appreciably enhanced by having been an inventor.

Given this view of invention, newness has sales value in itself. And this value persists, although perhaps with diminishing persuasiveness, even when there is no association between novelty and utility. Anyone with doubts need only notice the stentorian and repetitive emphasis which all advertising, even of the most stereotyped products, places upon what is new.

Only whiskey is an exception, and here advertising greatly emphasizes the novelty of the design of the bottle.

Additionally innovation in conjunction with advertising plays a vital role in stimulating the psychic obsolescence of goods and their replacement. This process, which is not without subtlety, has been most successful in the past in the automobile industry. But it also has a wide application to other consumer products and their packages. It consists in creating a visually new product and then, through advertising, persuading the consumer that this is the only valid image of the product. Although mechanical improvement, increased comfort or convenience or other technical advance may be claimed, these are not decisive for success. The important thing is that the change succeed in making the earlier version visually eccentric and that its possession and use, in consequence, reflect discredit on the person so owning and using it.

As usefulness becomes but one of the several factors justifying technical innovation—or, as with genital deodorants and plastic grass, usefulness must itself be imagined—the production and marketing of products of defective function or nonfunction becomes a normal feature of the economic system. The need for constant engineering of newness becomes (as in the case of automobiles) an endemic source of malfunction. Nothing can, in an engineering sense, be tried and true. It is too soon changed. And newness or seeming novelty, if it contributes to the effectiveness of consumer persuasion, serves better the purposes of the technostructure than reliability or workability.

Defective performance nonetheless provokes protest. Much of this protest misses the point. It is based on the belief that inutility and nonfunction are somehow aberrations in an otherwise perfect system—again a perverse manifestation of wickedness by corporations which know they should do better. It is useful to see, as this analysis shows, that the problem of inutility and nonfunction, so far from being adventitious and aberrant, is very much a part of the system.

Another systemic feature of innovation must be noted. Requiring, as it does, both capital and organization, technical innovation is largely confined to the planning system. So that is where resources committed to innovation are concentrated. Along with the importance of salable, as distinct from useful, qualities this explains the seemingly curious allocation of resources to innovation. Frivolous products of the planning system—those which seem to promise greater sexual opportunity, less obesity or some significant escape from the crypto-servant role of the housewife—will attract resources far beyond those accorded innovation to produce more efficient surface transportation or more comfortable or durable or less expensive housing.

4

In the management of the demand for public goods—purchases by the government—the role of technical innovation is, of course, crucial. It is also remarkably uncomplicated. Such innovation is highly organized and wholly deliberate. The purpose of a given innovation is, overtly, to render obsolete the preceding product and thus to create a demand for the newly developed product. Work then proceeds (or frequently will be already in process) on the next innovation with a view to rendering the new one obsolete and thus providing a market for the next product.

The procedure here described reaches its full perfection in the case of weapons and weapons systems, where the sequence of innovation and obsolescence has been fully systematized. Successive *generations* of aircraft, missiles, submarines, helicopters and main battle tanks are formally projected with the approximate dates in the future when the particular item will be made obsolete by further innovation and, accordingly, will require replacement. All associated with the process recognize its nature and know that the continued success of the industries concerned requires such obsolescence-producing innovation.

Support for the innovation—for the required research and development as well as for the later procurement—is further assured by interaction with the planning systems of other countries and most notably in the case of that of the United States with that of the Soviet Union. It is held that if the development that renders existing weapons obsolete does not proceed in the United States, it will proceed unilaterally in the Soviet Union. This, in turn, will accord an unsupportable military advantage to that country. It seems certain that the same argument as to the danger of unilateral development by the United States is made within the planning bureaucracy of the Soviet Union. The role of innovation in creating demand for public goods is thus reinforced by what amounts to tacit cooperation between the two industrial powers. The result is not only a powerful role for innovation in sustaining the demand for public goods but one that is largely immune to effective criticism. Any suggestion that innovation in weaponry is a device by which the technostructures of the weapons firms create the demand for their own products can be met with the answer that this may be true. But since development (it is held) cannot be controlled, there is no alternative that does not recklessly accord advantage to the other side. However costly or foolish, it must continue.

Additionally the intelligence which justifies the research and development comes through the public bureaucracy to which the technostructures of the weapons firms are symbiotically related. This intelligence—"what the Soviets are doing"—is adjusted, within limits, to need. Finally military secrecy can be invoked to exclude public and legislative intrusion into the relevant decisions. These can then be made by and within the organizations that have most to gain from the investment in innovation. The use of technical innovation in the management of demand for major public goods is, in all respects, the ultimate achievement in producer sovereignty. It is not surprising that here, at least, not even the most devout defend the traditional model of citizen or consumer sovereignty.

5

It is in light of the foregoing reality that public attitudes toward technical innovation have been altering. The neoclassical pedagogy still powerfully affirms its merit. But this instrumental conditioning is no longer fully effective in face of circumstance. And circumstance—the adverse experience with innovation—is, we see, firmly a part of the economic system. The technostructure in pursuit of sales exploits the public faith in what is new at the expense of what works. There is innovation that serves only to make a predecessor product visually obsolete. This too is advantageous. Innovation, even if serviceable, is irrationally distributed. It is concentrated often on unimportant things that are the product of strong organization, and it is slight for important things, the product of weak organization. And in relation to public needs—weapons in particular—the role of innovation is remarkably disquieting. In consequence the merit of innovation, both in private and public products, ceases to be something that can be assumed. Rather it is something to be assessed. To the means for making this assessment we will also return.

XVI

The Sources of Public Policy:
A Summary

Since the 1969 vote that reduced the fixed depletion allowance for the first time in 45 years, and as a result of mounting pressure on the industry to improve its record for preventing environmental pollution, oil company leaders have been concerned about the industry's poor public standing. The companies are planning to spend millions of dollars in television advertising to improve their image.

—The Washington Post
January 17, 1971

I was sitting with Helmut Strudel, president of Strudel Industries, at President Nixon's inauguration . . . The President said:

"Let each of us remember that America was built not by government, but by people—not by welfare, but by work—not by shirking responsibility, but by seeking responsibility."

Strudel began to perspire. "It sounds like he's not going to bail my company out of bankruptcy," he said worriedly.

"Don't be silly," I told Strudel. "When he speaks of people on welfare, the President's talking about the little guy who's freeloading on the government. He is not talking about companies that get large government subsidies."

—Art Buchwald, 1973

THE PLANNING SYSTEM, it will be evident, exists in the closest association with the state. The obvious core of this relationship is the large expenditure by the government for its products. This pays for the products of those corporations, most notably the large, specialized weapons firms, that exist by selling to the state. And it pays also for the technical development that sustains the cycle of innovation and obsolescence and thus the continuity of the demand. This same expenditure also contributes to a secure flow of purchasing power in the economic system as a whole and on terms highly favorable to the planning system.

The foregoing, however, by no means exhausts the claims of the planning system on the state. The planning system relies on the state for its large needs in qualified and educated manpower. This, as we shall presently see, is a matter of considerable social portent. The state not only educates those who accept and defend the values of the planning system. It also nurtures its critics—for there is no practical way of doing the one without the other.

Preceding chapters have mentioned other claims of the planning system on the state of which the reader may be reminded. The planning system requires supplementary investment by the state if it is to sell its products—as in highways if automobiles are to be sold. The planning system also needs support, either directly or indirectly through military expenditure, for technical development, much of it too expensive for the firm itself. Atomic power, computers, all modern air transports, satellite communication, numerous other products and services have their origins in such socialized technology. The government also supplies capital to the industries for which it is also the market; the large weapons firms use much plant and equipment belonging to the government and get their working capital in the form of progress payments. Finally the government is the lender of last resort for the occasional large corporation whose protective purposes are threatened by insufficient revenue or capital. The Lockheed Corporation, some stock exchange houses and the railroads have all been recent beneficiaries of this policy. The planning system has a powerful commitment to independence from the state except where public action is required. There is much to the increasingly common observation that the modern economy features socialism for the large corporation, free enterprise for the small.

Nor do the positive needs of the planning system from the state exhaust the list. It has many passive needs. The autonomy of the decision-making technostructure must be preserved. Resistance to particular kinds of growth or technical innovation must be neutralized. Large differences in income not related to merit or social utility must achieve social sanction. That the entrepreneur in the market system should have or expect the same level or security of income as the corporate executive of equivalent competence and energy in the planning system would be impractical and unthinkable. Similar if smaller differences in compensation must be accepted as between workers. The survival of the planning system in anything like its present form depends on its influence with and control of the state.

2

The public influence of the planning system requires, first of all, a secure public belief in the importance of the things that it does. Since it produces goods and services, this means that there must be a profound public conviction as to the importance of such goods and services. Much of this, as will also by now be clear, is a by-product of management of the consumer. That involves a large expenditure of art and money in newspapers, magazines, on billboards and, above all, on radio and television. No other form of communication is even fractionally so ubiquitous. All forms of consumer persuasion affirm that the consumption of goods is the greatest source of pleasure, the highest measure of human achievement. They make consumption the foundation of human happiness. In Europe, until the seventeenth century, the church, through its canonical spokesmen, had a near monopoly of communication with the masses. The not surprising result was a singular prestige for religious institutions and a passionate concern for the proper performance of religious offices. A pervasive if somewhat less complete command of communication by the planning system on behalf of its products leads to a comparable modern prestige for economic institutions and the goods and services they provide.

The planning system, having prestige as a source of goods and services and thus as a source of public happiness, will have influence as a source of political suggestion. What expands automobile or detergent production, in the absence of a strong showing to the contrary, will be wise and important public policy. Three related factors contribute to this prestige and add to political influence. As affluence increases, goods become increasingly dispensable or even frivolous. But it has always been a prime tenet of the neoclassical model that wants do not diminish in urgency and hence goods do not diminish in importance with increased output. Suggestion to the contrary was long held to be unscientific; the urgency of wants of an individual or social group at one point of time cannot be compared with that either of the same or different individuals or groups at a different and later period of time. Though wealth may have increased in the interim, so also may the longing for goods that must be requited. In consequence the yearning of a person for refined deodorants at a later period may be as urgent as for bread at an earlier time. This doctrine protects the importance of goods and the prestige of their producers in face of increasing output. And persuasion helps to accord serious importance to frivolous wants. It makes the taste or crispy sensation of the imaginative breakfast food important. It makes the shade of a sheet seem significant and thus the new detergent that accomplishes it. It gives similar meaning to other meaningless products. Thus it helps to conceal the tendency, with increasing production, to increasing unimportance. In technical terms it adds usefully to the impression of a constant marginal utility of goods over an indefinite range of increased production.

Persuasion—advertising and public relations expertise—also allows of direct appeal to the public where support or acquiescence on matters of public policy is sought. A charge that a corporation is polluting water or air, dissipating a natural resource or merchandising an unsafe product brings all but automatically an advertising campaign affirming that company's total devotion to environmental benefaction, resource conservation and the public safety. This is usually considered an effective substitute for more costly substantive action.

Finally persuasion provides a decisive source of income for radio and television stations and networks and newspapers and magazines. This does not directly purchase support for the goals of the planning system by the broadcasters, newspapers and periodicals, for that would not be practical. Newspapers and television must, of necessity, employ people who resist organizational discipline—whose instinct is to artistic independence. The number of people with such instinct is always larger than imagined. Men always recognize it in themselves while suspecting conformity in others. A very large number of people wish to be considered masters of their own thoughts, resist efforts to control them and regard themselves as unique in their stubbornness. Also for the media a measure of nonconformity is commercially advantageous. Where the public interest diverges from that of the technostructure and the planning system, as increasingly it does, people will reward with their patronage the newspaper or television station which articulates the public view. Heretical truth, even when inconvenient, is vastly more interesting than the predictable communication that serves the interest of the planning system.

The power of the planning system in relation to the media lies not in forthright control of expression but in its capacity to identify its needs with what, in public policy, seems basic and reputable. Thus, while interesting deviation has no difficulty finding a voice, the needs of the planning system are the norm to which discussion eventually returns. "Men with power have an extraordinary capacity to convince themselves that what they want to do coincides with what society needs done for its [own] good."[1] And this is the norm to which editors, pub-

---

[1] Raymond Vernon, "The Multinational Enterprise: Power Versus Sovereignty," *Foreign Affairs*, Vol. 49, No. 4 (July 1971), p. 746.

lishers and broadcasters, in the absence of thought to the contrary, also automatically repair. This is not made more difficult by the fact that the needs of the technostructure in general are also those of the larger publishers and broadcasters in their corporate incarnation.

3

The second objective source of public power is bureaucratic symbiosis. This has been sufficiently stressed. The market system approaches the government through the legislature. This relationship, though highly visible, is with the branch of government which has been declining in relative importance. The technostructure and the planning system have their relationship with the public bureaucracy. This association is far more discreet; it is also with the branch of government which, as public tasks become more complex, is strongly ascendant.

It is also an association that is uniquely available to the organization. An individual cannot effectively influence an organization; no single person can persuade, or bribe, the Pentagon on a major weapons contract, for, given the number of participants, such bribery would have to be by, and of, a battalion. Organization, by contrast, relates effectively to organization. The various specialists of the private bureaucracy work readily with their opposite numbers in the public bureaucracy pooling information for a jointly achieved decision. And, as earlier noted, rarely does the private technostructure meet a public bureaucracy without discovering some area in which there can be cooperation to mutual advantage. This is true even where the technostructure and the public bureaucracy are in a nominally adversary position.

Public regulatory bodies, it has long been observed, tend to become the captives of the firms that ostensibly they regulate. This is because the rewards of cooperation between the technostructure and the regulatory agencies normally outweigh those of conflict. The compliant regulatory body accedes to the needs of the technostructure; the latter supports or, in any case, does not oppose, the continued existence and needed budget expansion of the regulatory body. The aggressive regulatory authority, by contrast, invites public scrutiny of its needs. And, since its conflict is with the technostructure, it will be widely regarded as being in conflict with sound public policy. When it questions actions of the technostructure—the safety or quality of its products, the truth of its advertising—it is interfering with the natural prerogatives of private enterprise or hampering the growth of those innovations which, being the goals of the technostructure, are the foundations of sound public policy. Acquiescence, even if it risks criticism for being useless, may be better bureaucratic policy.

4

As the public bureaucracy has gained power in relation to the legislature, the latter has reacted to its own decline. The predictable reaction is protest and an effort to recapture lost authority. This has occurred. It is not necessarily the most powerful response. An appealing alternative for many legislators is to become allies of the public bureaucracy and thus, by association, to acquire some of its power. This has been the spectacularly successful choice of members of the Armed Services and Appropriations Committees. They derive power in Congress, patronage, public construction and weapons contracts for their constituencies and prestige in the community at large by identifying themselves fully with the interests of the military bureaucracies of which nominally they are the watchdogs. In doing so, they help ensure appropriate legislative action on the symbiotic needs of the planning system and the public bureaucracy. The legislative endorsement being in the name of the public, this also proclaims to the innocent the identification of interest between the planning system and the public.

A final source of political power for the planning system is organized labor. When there was conflict over the division of return, the antithesis between labor and capital was sharp. This, in turn, was reflected in political and legislative conflict. And for reasons deeply grounded in historical attitudes many people identified the public interest with the needs of the unions—with the toiling masses rather than the capitalist classes. In recent times, as we have seen, the conflict between labor and capital has been greatly eased by the ability of the technostructure to resolve conflict by conceding wage and other demands to the unions and passing the costs along in the price. This in turn makes possible a measure of psychic identification of the employee with the technostructure. The latter is no longer the implacable class enemy. At the same time the affirmative goals of the technostructure have become consonant with those of the union. A high rate of growth, which means steady employment, extensive access to overtime, perhaps even promotion, rewards the working force as well as the technostructure. So, accordingly, does the demand for goods that sustains such growth. This is powerfully true of government orders.

In the United States large defense budgets, a foreign policy that occasions such budgets, subsidized development of technology such as the supersonic transport, aid to temporarily ailing technostructures such as that of the Lockheed Corporation have in recent years had strong union support. This has been taken to be an aberration on the part of the leadership— the error of aged or obsolete men. This is unfair. It, in fact, reflects an exact assessment of the economic interest of the workers directly affected. The reflection of this in the legislature adds further to the public influence of the technostructure.

5

In assessing the public influence of the planning system, there is danger in being too specific. Its greatest source of such power is subjective. The technostructure consists of corporation executives, lawyers, scientists, engineers, economists, controllers, advertising and marketing men. It has allies and satellites in law firms, advertising agencies, business consulting firms, accounting firms, the business and engineering schools and elsewhere in the universities. Collectively these are the most prestigious members of the national community. They are generally the most affluent in a society that measures worth by affluence. Their view on public policy is the view that commands the solemn respect. And, with full allowance for the eccentric and often well-rewarded heresies hitherto mentioned, it is the view which reflects the needs of the planning system. It cannot be supposed to involve conflict with the public interest. What serves the technostructure—the protection of its autonomy of decision, the promotion of economic growth, the stabilization of aggregate demand, the acceptance of its claim to superior income, the provision of qualified manpower, the government services and investment that it requires, the other requisites of its success—*is* the public interest. Papal infallibility was powerfully served by the fact that the Holy Father defined error. The assurance that public policy will infallibly serve the technostructure and the planning system is similarly assisted by the ability of the technostructure to define the public interest. In recent years the existence of an accepted and strongly self-serving source of public policy has become generally recognized. It is called the Establishment. Popular instinct, as so often, identifies a basic truth.

More than the ability to say what is sound policy is involved. Involved also is the process by which men are selected to administer or execute policy. In a world of organization the values of organization are brought strongly to bear in selecting men for positions of public responsibility. Again the man who offers a divergent view—who departs from the Establishment position—is heard. But he is not thought fit for what is called real responsibility. That requires a man who accepts the goals of organization with a minimum of question, inner conflict or even ostensible thought. The planning system defines public policy in accordance with its own need. It also specifies the qualifications of those who carry forward the policy.

None of this is to say that the power of the planning system in defining and executing policy is plenary. With development its purposes tend to be increasingly at odds with those of the public. Increasingly effective persuasion is required. And, on occasion, this breaks down. Thus, during the decade of the sixties, the planning system and the associated public bureaucracy accepted the Vietnam war as they had long accepted the policy of resisting communism by military and other means in the less developed countries. This policy, despite great effort, could not be sold to the public. In more recent times similar though lesser fissures have opened between the public and the planning system on matters of domestic policy—notably on the environment, taxation, public support to such technological innovations as the SST, even on highway building. These fissures in the future will become wider. An identification of power is not an allegation of absolute power.

XVII

The Transnational System

[ITT] is constantly at work around the clock—in 67 nations on six continents . . . and quite literally from the bottom of the sea to the moon . . .
—Annual Report of ITT, 1968

THE PLANNING SYSTEM accommodates the state to its needs. What serves its protective and affirmative purposes becomes sound public policy. But the planning system also transcends the national state to create an international planning community. This is a logical and in many ways inevitable extension of the operations of the planning system as they are revealed within national boundaries. As domestic planning attacks the uncertainties of the domestic market economy, the international system deals with those uncertainties peculiarly associated with international trade. This system has developed in spectacular fashion since World War II. It can only be understood as an extension of the tendencies here described.

The planning system sustains a sizable priestly class which, for modest compensation, is available for industry meetings, customer and investor indoctrination, sales conventions, executive seminars and other corporate rites, where it combines a slightly extravagant didactic skill with a superficial aspect of deep thought. In recent times the favored topic of all such artisans has been the deeper meaning of the multinational corporation.

They picture the multinational corporation—IBM, General Motors, Nestlé—as a major discontinuity in the general process of economic development, something very different that, depending on the preference of the speaker and the need of the audience, is fraught with breathtaking potential for good or, more often, for evil. Standing astride of international boundaries, it is an assault on political sovereignty. Perhaps it is making obsolete the national state. And, since many of the corporations are American in origin, it is singularly a manifestation of American energy, enterprise and power. In the more advanced rhetoric it is the modern face of American capitalist imperialism. Not surprisingly, since much of this oratory is divorced from thought, most of this expression is wrong. The reality is less dramatic but not perhaps less important.

2

Old-fashioned international trade, as still described with intricacy and sophistication by the neoclassical model, is a system of exchange in which the market remains peculiarly powerful. The firm that engages in such trade is wholly subordinate to an impersonally determined demand and price. The market control by the firm being nonexistent, the uncertainty and risk of international trade is exceptionally high.

Specifically a firm in one country consigns its product, usually through intermediaries, for sale in another country. On leaving the country of origin, the product leaves entirely the influence of the original producer. The latter retains no power over the price at which the product is sold; he is not imagined to influence the preferences of the foreign consumer. The price of the product is subject to the competitive and uncontrolled consequences of domestic production in the country in which it is sold and of competitive imports from other foreign sources. Tariff changes in the recipient country and fluctuations in exchange rates add further uncertainty—both peculiar to foreign trade. If the firm is buying rather than selling—procuring, for example, an important raw material—classical international trade theory depicts a similar uncertainty.

Changing costs of supply, changes in the competitive demand from other countries, changes in exchange rates and (on occasion) changes in export taxes are added hazards. In this procurement the firm is also subject to the sole authority of the market.

The technostructure, we have seen, seeks as a major part of its protective purposes to bring its prices and its major costs under control and to ensure demand and supply at these prices and costs. And it has indispensable needs that must be supplied by the state. All this, in the aggregate, comprises its planning. Under the traditional system of international trade, that celebrated by the neoclassical model, all planning would have to stop at the water's edge. Firms might have the requisite domestic power. When they came to do business abroad, they would again be faced with all the uncertainties associated with the market system. To these would be added the extra uncertainties uniquely associated with international trade.

Nor would this be all. In addition to the uncertainties from buying and selling abroad, imports from other countries could prejudice or destroy control and resulting planning in the home country. This control—for automobiles, electronic apparatus, pharmaceuticals, whatever—consists in setting prices in accordance with the protective and affirmative needs of all participants and then forswearing the price competition that would be damaging to all. Foreign products—automobiles from Germany or Japan, electrical goods from Japan or Holland, drugs from Germany—would, in accordance with traditional theory, be sold at prices impersonally determined by the market. Supply would be governed by profit in the supplying country. Domestic producers would have to cut prices as necessary to meet the competition. Prices of these imported goods (being uncontrolled) would, on occasion, be further reduced to dispose of adventitiously large supplies. Thus the price equilibrium which characterizes the domestic oligopoly would effectively be impossible. To preserve control, firms would have to appeal for tariffs and quotas to insulate themselves from international competition. This would lead to similar action by other countries. Production would be confined to national markets. Where the retaliatory action would confine domestic firms to a relatively small domestic market as in the case of Belgium, Holland or even of Britain or France, growth —the principal affirmative purpose of the firm—would be very closely circumscribed.

3

The function of the multinational corporation is now clear. It is, simply, the accommodation of the technostructure to the peculiar uncertainties of international trade. It transcends the market internationally as it does nationally. It accomplishes over a world of multiple national sovereignties what it first accomplishes within any one. It minimizes the need for tariffs, quotas and embargoes to reduce uncertainty in national markets. And, needless to say, it is not peculiarly American. It is the common accommodation of all nonsocialist planning, whatever the country of origin, to the special problem of international trade.

By re-creating itself in other countries the technostructure, in effect, follows its product to those countries. In so doing, it enters into the same understanding on prices with the other market participants in the foreign country that it has on its home turf. And the reciprocal movement of foreign firms into its own territory eliminates the hazards of price-cutting and allows of the same control there. General Motors, operating through Opel in Germany, becomes part of the general oligopolistic convention that proscribes price competition in Germany. Volkswagen and Mercedes-Benz, selling in the United States, become similarly subject to the convention that outlaws destructive price competition in the United States.

Similarly, by following its product to Germany, General Motors can bring its persuasion to bear on the German consumer. And it can press its needs on the German government.

This, we have sufficiently seen, is indispensable. And, by following its product to the United States, Volkswagen is able to influence the responses of American consumers. And it can protect itself and obtain its needs in Washington. None of this would be possible were products consigned through agents in the manner of classical international trade.

Finally transnational operations ensure against other countries' obtaining a damaging cost advantage. This is a matter of great and increasing importance—especially to the United States. Within the United States there is no great danger that one manufacturer of automobiles, television sets, transistor radios, tape recorders, typewriters or other such products will have an insurmountable cost advantage over its rivals in the same industry. Unions tend to be industry-wide; the productivity of workers and access to capital, technology, qualified manpower and materials, if not the same for all firms, are not permanently and unmanageably different. As between countries, however, this cannot be assumed. In comparison with the United States labor costs in Japan or Germany may be durably lower, the productivity of labor may be permanently higher, capital costs may be lower and there may be continuing differences in the costs of other requisites of production. In consequence the costs of German or Japanese products may be permanently much lower than their American counterparts.

Given the foregoing differences—and given the reduced role of tariffs in the international planning system—the danger to the purely national corporation is evident. The oligopolistic convention under which prices are set serves well for firms that have roughly similar cost functions. But a foreign firm with sharply and durably lower costs might well insist on prices much below those that the domestic corporation finds advantageous. Movements in exchange rates, an occurrence of increasing frequency, add to the hazard. Both the protective and affirmative purposes of the corporation are thus in jeopardy.

With transnational operations this danger is elided. The

transnational corporation can produce or arrange production where costs are lowest. This advantage has been extensively and increasingly exploited, especially by American-based corporations, in recent years. Automobiles for all the major American producers are being built in Germany, England, Canada or Japan. Components for models being produced in the United States are assigned, routinely, to the lowest-cost foreign plant. And a very wide range of other products, notably electronic and other technical equipment, is produced to American specifications under American brands in Japan, Taiwan and Hong Kong. The threat of lower-cost foreign production is thus effectively countered by being (or using) the lowest-cost foreign producer.

In such fashion does the planning system transcend national boundaries and become transnational. "Multinational corporations are a substitute for the market as a method of organizing international exchange. [They are] . . . 'islands of conscious power in an ocean of unconscious cooperation.' "[1] That they have grown rapidly in the last quarter-century is not surprising. When we understand the nature of the planning system, we see how naturally they fit its purposes in international operations.

4

Not only is the multinational corporation a precise answer to need, it is an answer that greatly favors the most developed technostructure. Classical international trade is possible for any firm, large or small; it need only find someone to whom it can consign its product or, in actual practice, an intermediary who does so. Transnational operations require organization; they become increasingly feasible the larger the firm. The large

[1] Stephen Hymer, "The Efficiency (Contradictions) of Multinational Corporations," *The American Economic Review,* Papers and Proceedings, Vol. 60, No. 2 (May 1970), p. 441. The interior quotation is from a much earlier observation, dating to the nineteen-thirties, of D. H. Robertson. Since it was made, the islands, it could be said, have become continents.

firm has or can obtain the financial resources to establish operations or acquire firms in other countries and the managers, scientists, engineers and other specialized talent to re-create itself abroad. The larger the firm, the greater this advantage. IBM has the financial, managerial and technical resources to re-create itself in other countries. No foreign firm has the resources to do likewise in the United States in full-scale competition with IBM. Domestic development, we have seen, favors the highly developed technostructure. Transnational growth favors it even more.

Here, it follows, is the explanation of the eminence of American corporations in transnational operations. It is not because they are American. Where foreign firms have developed large and powerful technostructures, as in the case of Phillips, Shell, Unilever, Nestlé, Volkswagen, they have exploited transnational operations as vigorously as any American firm.[2] But the United States, befitting its higher level of industrial development, has the most advanced planning system. Accordingly it has far more corporations that are prepared for transnational operations than any other country. What has been called the American challenge is not American; it is the challenge of the modern planning system. This, because of size of country, absence of adverse feudal tradition, legal system, geography, resources and much else, has reached its highest development in the United States.

However another and opposite influence also explains the recent expansion of the transnational operations of American firms. As noted, the development of the modern economic system is highly unequal; this has been especially true in the United States where the weapons industries have been an area of especially advanced development. This development has

[2] The Japanese have been somewhat reluctant to accept the logic of the multinational corporation. This, from their viewpoint, has not been unwise. As the newest and most rapidly expanding industrial country they have needed freedom to force their way into other markets. And as a highly efficient producer they have had less need than others to bring foreign firms into the oligopolistic equilibrium that suppresses price competition in the Japanese home market.

led to a heavy concentration of capital and technical skills in these industries. It has also, since this is an area where costs are passed on with peculiar ease, done something to raise the general level of wages. In the past the United States had higher real wages than other countries but better technology and capital equipment. Of late, American civilian industries, formerly important in export or domestic markets, have been at a disadvantage in *both* quality of capital and wage levels in relation to Germany and Japan, where the military development has been much less. American corporations have responded in accordance with the preceding analysis; they have extensively developed lower-cost production abroad. This has enabled them to retain their position not only in overseas markets but in the United States as well. The eminence of the American corporation in transnational operations is a manifestation of both the high development of the American economy and of weaknesses that are associated with the particular form of that development.

5

As noted in Chapter XIII, the corporation in the planning system, in pursuit of the protective purposes of the technostructure, seeks to secure its sources of important raw materials. This takes it into the less developed countries. And to a limited extent in the case of products such as gasoline, transistor radios and aspirin it finds markets in these countries. The base being low, the rate of growth in operations is somewhat greater here than in North America and Europe. However the transnational system is primarily a relationship between the developed countries. It is here that the important markets exist; it is here, accordingly, that the protective equilibrium of the planning system must be established. Accordingly it is in the developed countries—Canada and those in Europe—that urgent questions are asked concerning the cultural impact of the

American multinational corporation and its political consequences for national sovereignty.

These matters are greatly misunderstood. The fear of cultural imperialism confuses the cultural impact of the American
corporation with that of the corporation in general. That the
planning system, in pursuit of its purposes, has a profound
influence on the national community in which it operates we
have seen. It conducts a powerful propaganda on behalf of its
products. Thus it shapes attitudes on behalf of goods in general
and makes these central to life. It imposes its intellectual discipline on those associated with its organizations. Its purposes
and values become the purposes and values of the community
and the state. But this cultural effect is not peculiarly American. This is imagined only because so many of the multinational corporations have their origins in the United States. As
a moment's thought will affirm, the cultural impact of equally
powerful technostructures originating in other countries is not
different. Were Belgium or Germany the headquarters of the
largest number of multinational corporations, the values, intellectual disciplines and pressures on the state that are now associated with American corporations would be associated with
Belgian or German corporations. The solemn and well-
regarded scholars would be speaking of Belgian or German
cultural imperialism. Though Canadians and Frenchmen think
much about the matter, it has never been shown that a Canadian who works in any capacity for General Motors in
Canada is subject to influences that are culturally distinct
from those encountered by a man who works for Massey-
Ferguson or International Nickel. Nor is a Frenchman who is
employed by Simca and thus by Chrysler in France less
French than one employed by Renault. The most notable feature of the modern corporation and thus of the planning system is the uniformity of its cultural impact, regardless of its
national origin. Its hotels, automobiles, service stations, airlines are much alike not because they are American but because all are the products of great organizations. Those who

The Transnational System   187

react to the American multinational corporation are reacting, in fact, to the cultural power of the planning system whatever its national identity.

The relation of the multinational firm to the sovereignty of national governments also now becomes clear. It greatly impairs such sovereignty. But that is not because of its transnational character; it is because the impairment of sovereignty —the accommodation of the state to the purposes and needs of the corporate technostructure—is the very essence of the operations of the planning system. The foreign firm comes into a country and impairs the sovereignty of the state. The domestic firms of similar scale and organization that are already there are already doing so. The foreign firm is more visible. So, accordingly, is its attack on the sovereignty of the state. When the French government helps Ford to locate in France, no Frenchman fails to notice. When it responds similarly to the needs of Renault or Citroën, no one pays much heed. But no one concerned with reality should be in doubt. The multinational firm invades the sovereignty of the state not because it is foreign but because that is the tendency of the planning system. The modern state, we may remind ourselves once more, is not the executive committee of the bourgeoisie, but it *is* more nearly the executive committee of the technostructure.

The multinational firm, although this is not a matter of great moment, will frequently be more restrained in its relation to governments, more scrupulous in the observance of the law, more cautious in its public persuasion than a domestically based enterprise. The press, politicians and people of any country will usually prefer a foreign malefactor to a domestic one. Recognizing this, the multinational firm will be exceptionally tactful in bringing its influence to bear on a foreign government. Where, as in the case of Italy, to cite only an example, tax evasion, bribery and the suborning of public officials are part of the operational routine, the firm will ordinarily seek a subsidiary with a strong domestic identity. That firm, in turn, will take responsibility for the requisite or conventional dishonesties in which, also, it will be technically accomplished.

The political case against the multinational firm regularly emphasizes that its wage and price policies are determined in its home country. In consequence workers and consumers are subject to a foreign authority to which they have no access and over which they have no influence. This, it will be seen, also misapprehends the reality. The multinational firm enters the country to become part of the wage- and price-making process of that country—to protect itself from the wage advantage of domestic producers or to be part of the price equilibrium that ensures it against disastrous price competition. Its protective and affirmative purposes are the same as those of the domestic firms. It does not, in consequence, exercise an independent authority on either wages or prices. The exercise of such independent authority, with its damage to common protective purposes, is precisely what its intrusion is designed to avoid.

6

Since World War II, there has been a marked retreat from what was once called economic nationalism. The visible manifestation has been the General Agreement on Tariffs and Trade (GATT), the European Common Market (EEC), the European Free Trade Area (EFTA), the several efforts between trading countries to reduce tariffs, restrict the use of quotas and, hopefully, forestall competitive currency depreciation. These developments have been attributed to the increasing economic enlightenment of the industrially advanced countries. In fact—a matter that will no longer have any large element of surprise—they reflect the needs of the planning systems of the participating countries. In the earlier circumstances of international trade, tariffs and other restraints protected the market system of one country from the advantages, comparative or absolute, of other countries. There was no substitute. No one controlled the supply from abroad or the price at which it was sold; an appropriate tariff was the only way of ensuring that it did not have a disastrous or even inconvenient effect on the domestic market price. With the development of the transnational system the intruding foreign firm does not reduce prices. That would be to break down the protective oligopolistic equilibrium into which it enters. If its costs are appreciably lower than those in the market it is invading, the firms that are subject to the intrusion have a remedy. They can produce in the country of the intruding firm. Or they can arrange for production there. In these circumstances tariffs are no longer necessary. By interfering with arrangements that can be better managed by the firms themselves, they could be a nuisance. We may be very certain that, had some tariffs been needed by the transnational system, the economic enlightenment which has led to their reduction or elimination would not have spread.

The transnational system and its needs are the clue to economic policy as between the developed countries. They similarly explain the resentments of the undeveloped lands, for the transnational system internationalizes the tendency to unequal development and to unequal income that has occurred domestically as between the planning and the market systems.

With the rise of the transnational system, capital, technology and qualified manpower are brought within the authority of a single organization. This authority extends across national frontiers. So does the ability to persuade customers and the community and to win needed support from the state. No such cosmic powers are available to the market system. The multinational corporations are in the developed countries; the undeveloped countries continue to conform to the market model. So the transnational system accentuates the inequality in development between the presently developed world and the rest.

The effect on income is similar. The planning system, we have seen, sells and buys at prices it controls. The transnational system internationalizes this power in the developed countries. The smaller firms of the undeveloped countries remain subject to the market—or to the market power of the transnational system. Both are beyond their control. Exploitation and self-exploitation, coupled with the barriers to the movement of workers across national frontiers, ensure that the resulting differences in income will persist and increase. Thus the transnational system also internationalizes the tendency to inequality as between the planning and the market systems. This, if one insists on the term, is the true shape of modern imperialism.

FOUR

THE TWO SYSTEMS

Instability and the Two Systems

[The Cold War] increases the demand for goods, helps sustain a high level of employment, accelerates technological progress and thus helps the country to raise its standard of living . . . We may thank the Russians for helping make capitalism in the United States work better than ever.

—Sumner Slichter, 1949

THE STRUCTURAL and operational outlines of the two systems are now in view. The planning system strongly dominates its environment; the market system accommodates itself to forces that it does not control. Development and income accord with this difference: in one case they are great, in the other much less. The difference in development, it must be noted, is when judged by consumer and public need. In the case of the planning system need is not only well served but extensively created; in the case of the market system need is frequently ill-served.

We come now to the final difference between the two systems. By itself the market system, like the classical combination of competitive firms and small-scale monopoly of which it is the modern prototype, is broadly stable. Fluctuations downward in output and employment or upward in prices do occur, but they are self-limiting and, eventually, self-correcting. The planning system, in the absence of state intervention, is inherently unstable. It is subject to recession or depression which is not self-limiting but which can become cumulative. And it is subject to inflation which is also persistent, not self-correcting. The consequences of recession and inflation in the planning system then overflow with profound and damaging effect on the market system. The latter suffers more from recession than does the planning system wherein the instability originates. We first examine the problem of downward instability or recession, then of inflation.

2

Downward instability—recession—in all normal peacetime circumstances is caused, in the first instance, by an insufficiency of effective demand, i.e., effectively used purchasing power. More goods and services are available or could be produced with available manpower and plant than there is demand for them. The planning system is prone to such a disparity.

Production, as a moment's thought will suggest, provides the income which buys the goods and services produced—each sale returns to someone the proceeds which, if spent, would provide the wherewithal for making the equivalent purchase. With that part of the proceeds which is, in fact, spent—which is turned to the consumption or further production requirements of the recipient—there is no problem. It is with savings that the danger of discrepancy arises. These must be invested and thus spent (or be offset by someone else's spending) or else there will be a deficiency in purchasing power. If there is such a shortfall, goods will remain on shelves; orders will fall; production will decrease; unemployment will increase.[1] Thus a recession.

In the case of the market system the danger of such a deficiency in demand is limited. Firms in the market system are numerous and small, and income is widely distributed in comparatively small amounts. The propensity to spend from this income is high—as it accrues, it is strongly exposed to the urgent consumption and production needs of the recipient. If there are savings, these will be deposited and made available

[1] This is a very plain model—no refinement, no qualification. It is, however, the one that is the core of all modern discussion of stabilization policy.

for lending. And if they are sufficiently abundant, there is at least a chance that this will be at interest rates and on terms that will encourage the other and always needy firms of the market system to use them.

Moreover, if an effort to increase savings, which would be offset by increased spending or investment, should reduce total demand, the market system has certain back-up mechanisms which limit the damage. Prices fall, but the self-employed entrepreneur, though his income suffers, does not become unemployed. Nor do the members of his family. Nor, in the frequent case, do his workers, who are rewarded instead with a cut in their wages. Meanwhile the reduction in income immediately reduces the ability of the entrepreneur to save—and thus ensures that more of what he receives is spent. And the lower prices for products and services attract the custom and increase the purchases of those who are living on fixed incomes or who are spending from past accumulation. So an equilibrium in which demand covers supply is likely to be reestablished at lower prices. Output does not decline; there is no increase in unemployment.

None of this works out quite so well in practice. Prices and wages in the market system are always more rigid than here suggested. Output might fall; unemployment might rise. But in the market system savings do accrue widely in small amounts, and are likely to be used. The small entrepreneur does reduce his income and remains employed. His savings do fall when this happens. In the aggregate the tendency in the market system is to stability.

3

In the planning system circumstances are very different: the factors making it uncertain that savings will be spent are much more powerful; the corrective or back-up mechanisms do not exist.

With the rise of the large corporation saving, as we have

seen, is largely from business earnings. In 1972 these (retained business earnings) were $124 billion as compared with personal savings of $55 billion.[2] Much the largest share of the $124 billion was by the large corporations that comprise the planning system. This saving is well removed from the temptations of consumer expenditure. There is agreement in both socialist and nonsocialist planning systems that the choice as between saving for investment and spending for consumption must not be made by individuals, for then much too much is spent, much too little is available for investment.

As the savings decisions of the planning system are made by a comparatively small number of large corporations, so also are the decisions to invest. Large magnitudes are involved. And there is no mechanism by which the two sets of planning decisions—the one involving savings and the other involving investment—are matched. Not even the most ardent defender of the neoclassical system imagines that the market any longer serves—that interest rates fall as necessary to discourage excessive savings and to encourage insufficient investment so as to keep the two equal. Accordingly intentions to save can easily exceed intentions to invest. In consequence there can be a deficiency in demand. As this deficiency reduces output and renders a plant idle, it will reduce investment further and thus aggravate further the deficiency in demand. This process will continue until declining profits have a more than offsetting influence on savings.

Additionally the planning system, unlike the market system, returns income to individuals in large, coarse amounts. These large incomes are better insulated from the pressures of consumption than the modest amounts distributed by the market system. They add to the total volume of income where there is an option between spending and saving and thus between ultimate expenditure and nonexpenditure.

Spending on the products of the planning system is exten-

[2] Economic Report of the President, 1973. (Figures subject to minor revision.)

sively the result of persuasion. However effective this management, the resulting consumption is less reliable than that derived from individual discovery based on pressing need for food, shelter, medicine or clothing. When incomes in the planning system fall, people can cut their consumption far more readily than in the market system.

Finally in the planning system the back-up mechanisms which serve to stabilize the market system—and to limit the tendency for a downward movement in demand to become cumulative—are inoperative. Prices, being subject to the control of the firm, do not fall. And wages, being subject to the authority of the unions, cannot be reduced. Accordingly, when demand falls, there is no offsetting effect of added sales from lower prices. And there is no chance that the effect of lower wages will be offset by increased employment. The entire impact of the reduction in demand is on output and employment.

The instability to which the planning system is subject has an adverse effect on both the protective and affirmative purposes of the technostructure. But this instability has also a powerfully adverse effect on the market system. When demand in the planning system falls, demand for the products and services of the market system is reduced. Since there is no protective control, prices, entrepreneurial incomes and some wages fall. Hardship for the small businessman or farmer is severe. While the market system can contain movements in demand arising from within itself, it is extremely vulnerable to adversity emanating from the planning system.

4

Recognition that the modern economy was subject to severe downward instability, and that this was neither self-limiting nor self-correcting, came in the decade of the thirties. This was the Keynesian Revolution. A word of summary will be sufficient. As originally envisaged, the government would intervene with increased civilian expenditure of general public benefit, not covered by taxes, to offset the deficiency of aggregate demand. But, following World War II, the Keynesian Revolution was, in effect, absorbed by the planning system. Thereafter government policy reflected closely the planning system's needs. Public expenditures were set at a permanently high level and extensively concentrated on military and other technical artifacts or on military or industrial development. These expenditures provided direct support to the planning system and were immune to congressional or other attack, for they served the higher purposes of national security or other national interest. Specific stabilization was then accomplished by tax adjustment—by the personal and corporate income taxes which rose and fell more than proportionately as income and therewith demand rose and fell and which were subject, as needed, to specific change.

How well this development was accommodated to the needs of the planning system will be evident. Instability originated within the planning system. But it had painful effect on prices and incomes in the market system and particularly on workers. From these last victims came broad political support for remedial action. The remedial action was highly serviceable to the protective and affirmative purposes of the technostructure, both of which were jeopardized by downward instability in the system, or recession. And the remedial action, which was pivoted on public expenditures that purchased the products, supported the technical needs or facilitated (as in the case of highways) the sales of the planning system. These expenditures, especially those for weapons, were then made secure by associating them not with economic policy but with the much more sacrosanct purpose of national security.

Very little of the change just described became part of the accepted body of neoclassical or neo-Keynesian analysis and instruction. The downward instability of the economy was not identified very closely with the rise of the great corporation. Keynesian economics was regarded as a discovery, not as an accommodation to change. And the role of the planning system in accommodating the Keynesian policy to its own needs was kept almost completely out of sight. Three conventions of neo-Keynesian and neoclassical economics served to disguise the reality. A word of reiteration is in order.

First, it was taken for granted that prevention of unemployment—together with a satisfactory rate of growth in the economy—was of such importance as to override all questions of method or benefit. Were unemployment above a certain level, the economic policy was a failure. If unemployment was at a tolerable minimum, one did not inquire how this was achieved. With unemployment, as with a coronary thrombosis, one concerned himself only with the efficacy of the cure.

Second, in what was doubtless the most direct manifestation of social conditioning on economic instruction, it was felt to be improper, possibly mischievous and certainly unscientific to suggest that military spending was functionally integrated with economics. Radicals might say so. Others might allow themselves the luxury of belief. But the responsible scholar in his pedagogy denied the possibility. "There is nothing about [government] spending on jet bombers, intercontinental missiles and moon rockets that leads to a larger multiplier support of the economy than would other kinds of expenditure (as on pollution control, poverty relief and urban blight) . . . America's potential and actual growth rate, far from depending upon war preparations, would be markedly *increased* by an end to the cold war."[3]

Finally there was, to repeat, the convention that separates microeconomics from macroeconomics. Microeconomics made the firm safely subordinate to the market. No power was left over, except as an aberration, to influence or guide the state. And another set of scholars in another set of courses considered the effect of public expenditure and taxation (as well as monetary policy) on the economy as a whole. It was no part of their business to see how the corporation influenced the

[3] Paul A. Samuelson, *Economics,* 8th ed. (New York: McGraw-Hill, 1970), p. 804. Emphasis in original.

nature of public expenditure. So the accommodation of macroeconomic policy to the needs of the planning system was (and remains) effectively insulated from economic scrutiny.[4]

With the rise of the planning system the economy became systematically subject to downward instability—to recessions. The same development also made it systemically subject to inflation. Recognition of the tendency to inflation was much more reluctant than in the case of recession. This was partly because both the tendency and the remedies were far more difficult to reconcile with neoclassical orthodoxy. To these matters we now turn.

[4] But not completely. In his book, *The Three Worlds of Economics* (New Haven: Yale University Press, 1971), Professor Lloyd G. Reynolds argues that the division between micro- and macroeconomic problems is being used, in effect, as a screen for the reality.

"Most of the policy issues which have emerged during the past generation—poverty and income distribution, deterioration of urban areas, equality of educational and occupational opportunity, overpopulation, environmental control, removing the blemishes of the market economy—are micro problems. Analysis of collective decision making also requires micro reasoning, whether of a normative character (cost-benefit analysis), or an explanatory character (models of citizens and legislators). Those who still insist that macroeconomics should receive major weight all the way from the principles course through graduate school are practicing a conventional wisdom just as obsolete as that of their predecessors in the 1920's . . ." (p. 310).

XIX

Inflation and the Two Systems

THE MARKET SYSTEM can suffer from inflation—from an enduring increase in prices. But such inflation is not inherent in the system, and it is rather readily remedied.

The individual producer in the market system takes his instruction from prices that he does not control. If there is a sufficiently strong demand, this will pull up his prices. But the source of such demand, as a practical matter, will be either lending by banks (or other sources of credit) in excess of what is being saved or spending by government in excess of what is being taxed. There is a remedy for both: Credit can be tightened up by the central bank; the government can tax more or spend less or both. If the remedy is sufficiently pressed, the lowered demand will cease to lift prices. The producer does not have the power to resist. Prices will be stable—or fall.

Unions can make things more difficult. But in much of the market system they do not exist. The work is done by self-employed entrepreneurs or by a few unorganized employees. Where unions do exist, producers do not control prices. If demand is being restricted and prices, in consequence, are falling, union pressure for higher pay will be resisted.

In a number of industries in the market system, notably the clothing and construction trades, numerous small businessmen deal with one or a few strong unions. Here wage increases do shove up prices; an industry-wide or community-wide wage increase forces the employers, in effect, to agree on a price increase. Thus the union has the price-making power that the employers lack. Even here, however, much will depend on whether the unions in question are pressed by increasing living costs and by the example of wages in other trades. If prices and wages generally are stable, unions in this part of the market system will not be under pressure to seek higher pay. And since there is an immediate effect on prices, they will have to reckon with the effect on sales of their product and therewith on employment among their members.

In the planning system the position is very different. Here the firm has power over its prices. An important purpose of this power, we have seen, is to allow wage costs to be passed on to the public. This serves the protective purposes of the technostructure—it ensures that wage increases will not bring a damaging reduction in earnings. Other factors make it less than likely that wage demands will be resisted.

When profits are being maximized and when they are being received by the entrepreneur himself, wage increases reduce returns. And they reduce them for the man who controls the decision as to whether or not the increases will be granted. It adds to the distaste for a wage increase if one has to pay it. Resistance is to be expected. When power is possessed by the technostructure, the decision is taken by men who do not themselves pay the cost. And since profits are not being maximized—since there is, in effect, an unused possibility for monopoly gain—it will frequently happen that by raising prices profits can be kept at their former level.

To be sure, higher prices are inimical to growth. But in the short run growth will frequently be better served by uninterrupted production than by a nasty strike. In the longer run the support that is accorded by the state to aggregate demand is, we have now seen, an extension of the power of the planning system. It will thus be assumed, and correctly, that, should demand be insufficient to clear markets at the higher prices, the state will sooner or later take action to make up the deficiency.

All of this means that, in the planning system, the normal tendency is to accede to the wage claims of unions. The ceremonial insult that graces the collective bargaining process slightly disguises but does not alter this circumstance. Left to itself, and given its control of aggregate demand, the planning system will show a steady upward movement of wages and prices.

The existence of rival unions of varying power adds to this upward instability.[1,2] One union seeks naturally to improve on the settlements won by others. The union with the greatest determination and power thus sets the pattern for the others. As prices and living costs rise, unions must also seek provision in the settlement for prospective increases in living costs—or negotiate escalation clauses to cover such increases. This means larger increases in prices and leads in the next round to yet greater increases in wages. Thus the tendency of the planning system to a continuing and cumulative upward spiral in wages and prices.

In the last twenty-five years all industrial countries have brought the downward instability of the planning system under control. The specter of the cumulative downward spiral in wages, prices and production has receded. Instead industrial countries have uniformly found themselves contending with upward instability—with inflation. And, as with cumulative recession, this inflation has flowed over from the planning sys-

[1] My colleague, Gail Pierson, has studied, for the fifties and early sixties, comparative wage changes in strongly and weakly organized industries at different levels of employment. Her conclusion is that "union strength does make a difference; it significantly worsens the terms of the tradeoff between unemployment and inflation." "The Effect of Union Strength on the U.S. 'Phillips Curve,'" *The American Economic Review,* Vol. 58, No. 3, Pt. 1 (June 1968), p. 456 et seq.

[2] Yet more recent and direct verification of the relationship between corporate market power and wages as here outlined is in "Market Power and Wage Inflation," by Daniel S. Hamermesh, *The Southern Economic Journal,* Vol. 39, No. 2 (October 1972), p. 204 et seq.

tem to unsettle the wage and other cost structures in the market system and also in the public sector of the economy.[3]

2

The susceptibility of the economy to recession and to inflation trace, it will be evident, to the same causes—to the rise of the planning system and the associated appearance of the modern union. The history of attitudes on these two matters has, however, been very different. The notion of downward instability (in the economy as a whole) received fairly ready acceptance. Not so the notion of systemic inflation. The case for an inherent downward instability was influentially made in Keynes's *General Theory*[4] in 1936. Within a decade the belief that the modern economy was subject to a deficiency in demand—and that offsetting government action would be required—was close to becoming the new orthodoxy. It was accepted into law with the Employment Act of 1946. By then a business organization, the Committee for Economic Development, had come into existence for, along with lesser purposes, propagating the new faith. Thanks primarily to the powerful and unfettered initiative of Professor Paul Samuelson it was soon to be standard in economic pedagogy. That the economy might be unstable in an upward direction was accorded no such acceptance. In appreciable measure it still lacks it.

The most important reason for the difference is that steps to correct the downward spiral could much more readily be reconciled with the basic principles of the neoclassical system. There was a deficiency in demand; the state offset this deficiency. This, in effect, propped up the system. Everything then func-

[3] Parts of the economy where, characteristically, wage increases are not offset by productivity gains. In consequence the effect of such cost increases on public budgets or on prices (where they could be passed on) has frequently been greater than in the planning system. This has caused scholars who preoccupy themselves with the surface of things to suggest that inflation originates in the market or public sectors of the economy.

[4] *The General Theory of Employment Interest and Money* (New York: Harcourt, Brace, 1936).

tioned very much as before but at a higher level of output, income and employment. The market remained as in the textbooks. The firm remained subordinate to the market, and thus to the individual. The neoclassical ideas escaped fundamental damage. So viewed, the Keynesian Revolution was a small revolution. Economists, like others, much prefer small revolutions to larger ones.

The problem of inflation was also thought, for a long while, to be a by-product of action against downward instability. In propping up the economy there was always danger of overdoing it—of providing more demand or stimulating more spending than the economic system could satisfy at current prices. Prices would then be bid up, as would also be labor or those kinds of labor in scarce supply. The economy, in the homely phrase of the cognoscenti, would be "overheated." Since the market was assumed to be unimpaired, however, a curtailment or reduction in public spending, taxes remaining the same, or an increase in taxes, expenditures remaining the same, or a curtailment of private spending from borrowed funds would stabilize or reduce aggregate demand. This would be reflected by the unimpaired market to the firm. Prices would cease to rise. And with the stable or lower prices (or lesser output) bidding for scarce labor would come to an end, wage increases would be resisted, and—since living costs would be stable—wage claims would be less ardently pressed. What cured a deficiency of demand when put in reverse would cure inflation. The transition from the propping up to the curtailment of demand would be relatively painless. In the words of one optimistic but well-regarded spokesman for the approved view, "In a gradual restoration of reasonable price stability, the rise in unemployment can be fairly small and a recession certainly need not be part of the prescription."[5,6]

[5] Arthur M. Okun, *The Political Economy of Prosperity* (Washington: The Brookings Institution, 1970), p. 101.
[6] As always there were notable dissenters. "The fatal flaw in the practice of the New [i.e., Keynesian] Economics stems from . . . The underlying notion . . . that there is an ideal target—full employment without inflation—that can

The foregoing conclusions, it may be noted, were compelled in considerable part by the need to protect the basic faith. Here, as in the case of profit maximization or the management of demand by producers, was one of those critical points that could not be conceded without irretrievable damage to neoclassical belief. For, if producers have the power to raise prices in response to wage claims and if the increase depends on the amount of the claim, the firm is no longer subordinate to the market. The neoclassical model is a chimera; extensive control of prices by producers—planning—must be conceded. And, if this planning requires the intervention of the state to fix wages and prices, the game is wholly up. One cannot have a market system in which there is massive wage- and price-fixing. A small revolution would become an inconveniently large one. Thus, and understandably, the resistance.

3

Two other factors delayed the acceptance of systemic inflation based on the interaction of wages and prices in the planning system. The division of labor as between economists was again sadly important. In the thirty-five years following the publication of Keynes's *General Theory* the continued growth of the great corporation was duly noted by economists, as also its power over its prices. A few came to accept its concern with growth as a primary goal. And the less militant character of labor relations was remarked—as also the tendency to settle disputes by passing the costs along in higher prices. But, again, these matters belonged to microeconomics or were the concern of specialists in industrial organization or labor relations. Macroeconomics, having to do with the management of aggregate demand, was as always in a different compartment,

be achieved simply by a proper adjustment in the government's aggregate expenditures and taxes . . . the assumption finds no support in experience. At no point in history, even temporarily, was the target achieved in any country where the New Economics was practiced." Melville J. Ulmer, *The Welfare State: USA* (Boston: Houghton Mifflin, 1969), p. 33.

the work of different scholars. Corporate and union power was not their preoccupation. The division of labor, with its implicit assumption that the two parts of the economy could be studied in isolation, fell at exactly the point where it best disguised the impact of corporate and union power. Given the damaging effect of that power on accepted and convenient belief, nothing could have been more agreeable.

Finally it was possible for the planning system to live in reasonable comfort with both inflation and the orthodox efforts to control it that assumed the continued pre-eminence of the market. These last could be painful, but the pain was concentrated on the market system. This the planning system could endure.

Planning is unquestionably easier with stable prices and international exchanges. But inflation does not impose changes on the firm that are wholly beyond its control. Rather inflation is a process which reflects, in part, the power of the firm. The upward movement in prices reflects that power and reflects, also, the ability to offset wage and other cost increases that are not fully controlled. In recent times a considerable number of economists, responding as before and with undoubted innocence to the tendency to identify sound economic policy with the needs or preferences of the planning system, have begun to argue the acceptability of continuing inflation. The question is only how much. This position has the further and very great advantage of avoiding government wage- and price-fixing and the frontal assault that this mounts on neoclassical belief. The market can be assumed to survive inflation but not price-fixing.

The planning system is also little damaged by the orthodox measures for attacking inflation—those that assume the continued pre-eminence of the market. There are, to remind, three ways by which demand can be reduced—reducing public expenditures, reducing private expenditures from borrowed funds and raising taxes. All rest lightly on the planning system. The public expenditures which buy the products of, or directly serve the needs of, the planning system cannot be much reduced. They have, as earlier noted, the sanction of higher national purpose—"you cannot gamble with the nation's security." In consequence, if expenditures are to be foregone, postponed or reduced to prevent inflation, it is upon outlays for welfare, housing, urban services, education and the like that the curtailment has principally to fall. Thus the initial impact of budget restraint is not on the planning system but on the civilian services of the public sector or those of the market system. It is not easy to reduce any public outlays. Still, if inflation is endemic, as it has been for the last quarter-century, the result will be evident. Pressure for curtailment in the civilian services of the state and against increases will be persistent. No similar pressure will operate against expenditures for the products of, or on behalf of the needs of, the planning system.

The second way of reducing demand is by increasing interest rates and otherwise reducing the supply of funds for lending and thus for investment spending. Here the differential impact on the two systems is forthright. It is central to the protective purposes of the technostructure—the protection of its autonomy—that it minimize its dependence on borrowed funds. Instead it relies extensively for capital on its own earnings; thus the overwhelming importance of business savings as the modern source of capital. The market system, in contrast, relies heavily on borrowed funds. Farmers, small retailers, small dealers are persistent borrowers. (States, cities and school districts are also heavy users of borrowed funds.) In housing and other construction this dependence is nearly total. Additionally the large corporation of the planning system, when it must borrow, is a favored client of the banks, insurance companies and investment bankers. If a conglomerate, it may even have a financial affiliate. The small retail borrower of the market system has no similar standing.

It follows that steps to raise the interest rates, restrict lending and thus reduce spending from borrowed funds have a radically different effect in the planning and in the market systems. In the planning system, firms, comparatively speaking, are little affected. In the market system they feel the full force of the policy. Recurring use of monetary policy means a recurrent restriction of the development of the market system as compared with the planning system. It has a similar restrictive effect on borrowing for public purposes, notably by states, localities and school districts. It has long been observed that in periods of monetary restraint—times of what is called tight money—there is severe complaint from farmers, the residential housing industry, merchants and other small businessmen. Big firms rarely object. This has commonly been attributed to the inferior moral tone of farmers and other small entrepreneurs—their lesser willingness to accept the salutary pain that is an indispensable feature of free enterprise. A deeper reason for their complaint now appears. The further effect of this policy in distorting development in favor of the planning system will also be clear.

The third way of restricting demand is by increasing taxes. Here too there is some difference in impact as between the two systems. The planning system has control over its prices as the market system does not. So it is able to pass higher taxes on goods and services and perhaps even higher taxes on income along to the public as the market system is not. And should taxes fall on profits, it is not the members of the technostructure but the owners who pay. In the market system, with rare exception, it is the man who runs the business who must pay.

4

Yet, while the orthodox measures against inflation are far more acceptable to the planning than to the market system, they have the further and unfortunate aspect that they do not work. Nor, given the structure and purposes of the planning system and its relation to the unions, is there any chance that they ever will. Meanwhile decades and even centuries of earlier conservative economic instruction have made stable prices—a coinage of stable value—a good thing. Rising prices remain unpopular—a symbol of inadequate or incompetent government. The consumer, even though his own money income is rising, feels, when he finds prices have gone up, that he has somehow been robbed. And inflation complicates the collective bargaining process. Wage gains are won and promptly eroded by price increases; the whole process to the worker comes to have an aspect of fraud. He gets something only to have it taken away. Among the other questions he asks is what good is a union?

Most important, the measures to restrain inflation, although they weigh lightly on the planning system, weigh painfully on the market system. The members of the market system have votes. And since the measures are ineffective, there is pain without the compensating advantage of price stability. Effective restrictions on aggregate demand—although their primary incidence is on the market system—also adversely affect economic growth. So they are in conflict with the affirmative purposes of the planning system. In consequence, the damage to neoclassical beliefs notwithstanding, overt government action on wages and prices becomes inescapable. Circumstances, as so often, ride ruthlessly over the most needed doctrine.

5

Wartime planning apart, the first efforts at wage and price control in the planning system were initiated in 1961. These—the setting up of guideposts limiting wage increases to average productivity gains and requiring companion stability in prices—held prices generally stable for the next several years and in face of a steady expansion in output and employment. The policy was not, however, central to the beliefs of the men administering it. By keeping it on a voluntary basis it was imagined, however illogically, that an underlying structure of free markets was being maintained intact. In the later sixties the pressure of wage claims increased, as did the pull of demand occasioned by the Vietnam conflict and the associated reluctance to add to the resistance to an unnecessary war by raising taxes. Paradoxically, as the need for control increased, the guideposts were abandoned. Those responsible for the guidepost policy returned to the universities to instruct students in the pre-existing orthodoxy. Inflation became more serious and overflowed strongly into the market system and on wage and salary costs in the public sector.

With the advent of the new Republican administration in 1969, the problem of inflation was attacked anew. The economists who then came to Washington were men of unquestioned piety in their commitment to the neoclassical faith. They strongly disavowed the by then ineffective efforts at wage and price restraint of their predecessors. The current inflation they firmly identified with insufficient application of orthodox remedies. They made especially clear their commitment to a much more severe monetary policy, meaning a much reduced expenditure from borrowed funds. They then awaited results with the excellent morale of men whose actions are positioned firmly on principle. Month by month for two years and a half and with a sincerity that was diluted only by the hope that their optimism might be infectious, they predicted a leveling off of prices with, at most, only a moderate increase in unemployment. For two years and a half prices rose without effective interruption. So did unemployment. In the market system house construction declined sharply in face of a housing shortage. Prices for farm products remained fairly stable. Farm costs, set generally in the planning system, rose relentlessly in relation to those prices. The effects of the policy were not confined to the market system. Tight money and the general reduction in demand created difficulties even for a number of the more vulnerable firms in the planning system. Conglomerates, as noted, pursue a high-risk strategy of growth involving the acquisition of other firms with borrowed funds. They suffered. The Penn Central, which had combined such acquisition with unusually incompetent management, went into receivership. In a uniquely visible demonstration of bureaucratic symbiosis the Lockheed Corporation saved itself from a similar fate by obtaining public underwriting of its debts. "We are going to continue to slow down the rate of inflation in the middle of an orderly expansion," President Nixon promised at the beginning of 1971, adding that ". . . we are going to do it by relying on free markets and strengthening them, not by suppressing them."[7] By the middle of the summer of 1971, the neoclassical faith of his economists was something that a President with an election only a year and a few months distant could no longer admire. Orthodoxy was jettisoned—as a student of this analysis would, one trusts, have predicted. The tight money policy was abandoned. Restraints on public expenditures were eased, and a budget deficit became official and admired policy. Virtually all wages and prices were frozen, and this was followed by steps to develop a system of wage and price control specifically applicable to the planning system.

The commitment to public wage and price regulation is not yet firm.[8] Since the policy has had no standing among reputable economists, virtually no thought has been given to the problems of administration. (Minute details of central bank administration are part of the established ritual of economic instruction in all sizable universities. Few or none have taught how prices or wages might be controlled.) For this reason, along with that of inherent complexity, the administration of price and wage control will for a long while be messy. Also men of established faith still defend the controls not as an

[7] Economic Report of the President, 1971.

[8] Since this was written (and last revised) the President's economists, finding that the policy of control had moderated the rate of inflation, jettisoned most of the controls. (The notion that if a policy is working, it should be abandoned reflects a novel approach to public policy.) There ensued, in early 1973, a renewed and even more serious inflation exaggerated by a diminished confidence in the prospective purchasing power of money, a general rush into goods and a strong bidding up of prices especially in the market sector. As this book goes finally to press, new controls have again been announced.

intrinsic feature of the system but as a temporary expedient. This will lead them, when there are controls, to urge a return to the free market, as it is righteously termed. And if their advice is heeded, renewed inflation will force renewed controls.

Fairness, nonetheless, requires one to commend President Nixon's contribution to economic understanding. The constitution of the neoclassical system cannot be reconciled with general wage and price control. The structure of the planning system presumes it. Thus the President has gone far to attest to the existence of the planning system. One cannot be sure that this book will accomplish more.

XX

The Economics of Anxiety: A Test

Never before in world history has a nation been so endowed with wealth and power, yet so plagued with doubt as to the proper use of that wealth and power.

—Walter P. Reuther
"Goals for America"

"The Income Revolution," touted in the 1950's as the graceful succumbing of inequality to economic growth, has not occurred.
—S. M. Miller and Pamela A. Roby
The Future of Inequality

THE ULTIMATE TEST of a set of economic ideas—a system, if the word be allowed—is whether it illuminates the anxieties of the time. Does it explain problems that people find urgent? Does it bear on the current criticisms of economic performance? Most important, perhaps, does it bear upon the issues of political debate, for these, though many have always preferred to believe otherwise, do not ignite spontaneously or emerge maliciously from the mouths of agitators to afflict the comfortable. The time has come to review the major anxieties of the time and see whether, in fact, they are explained by the ideas here offered.

Few will question that one of the complaints concerning the modern economy is irrational performance. By now the complaint lacks novelty. Some products of the private sector of the economy—automobiles and their fuel, drugs, cosmetics, intoxicants, packaging, exotically processed foods—are available in abundance. Other things of obvious importance—housing, medical care, mass transportation—are endemically in short supply.

The neoclassical system recognizes areas of deficient production. These are the industries characterized by monopoly and oligopoly. That this explanation is obsolete becomes evident when it is observed that the industries commonly associated with irrationally excessive production—the automobiles, drugs, cosmetics, intoxicants, containers and processed foods just mentioned—are the classical examples of oligopoly. For the rest the neoclassical system holds that the consumer distributes his purchases in response to inner will in such fashion that his satisfactions from outlays for different products and services are equal at the margin. This, in the private sector of the economy, determines what will be produced. If things of importance are not produced, it is because consumers understand badly their needs. If the distribution of productive resources seems insane, it is because consumers are insane. The economic system does not distinguish between the significant and the insignificant, the serious and the bizarre.

The explanation deriving from the present analysis is more readily reconciled with circumstance, common sense and the assumption that people are sane. The performance of the economy in relation to need is, indeed, unequal. That is because power to organize resources and to persuade consumers and the government as to need is unequally distributed between the market and the planning systems. To the unpersuaded, or to others when they cease to be mesmerized, the result is irrational. And the irrationality is enhanced by the superior capacity of the planning system to win support from the community and the state for the things that (in the manner of highways for automobiles) support its production and—as shown in the last two chapters—by the way in which the stabilization of the economy favors public expenditures on behalf of the planning system and discriminates against borrowing and investment by the market system. That there will be complaint that production is irrational is surely predictable when the view is through the present lens.

2

There has been specific complaint in recent times over the irrationality of the expenditures of government. Weapons, aircraft developed to a level of general nonperformance, moon travel, the space shuttle, atomic tests, industrial research and development, highways have ample access to public funds. Money for public needs of the highest importance or greatest public convenience—education, police, law courts, street sanitation, urban services of all kinds—is persistently in short supply. Some of these services in cities as distant from each other as New York, Los Angeles, Rome and Tokyo are so scarce as to make life presently unpleasant and potentially short. That in the United States we have somehow "got our priorities on public expenditure wrong" has become a cliché.

The neoclassical system is again unhelpful. The distribution of public expenditures remains in response to citizen will. Either the citizen is committed to his own discomfort and perhaps even his eventual extinction or there is, for the moment, some sui generis flaw in government which is producing the wrong use of public funds. The present analysis again makes the result predictable. There are some parts of the market system, most notably commercial agriculture, which are able to bring their power to bear on the government through the legislature. Otherwise the distribution of public resources reflects the power of the planning system over the state. Where that power is great, the services are ample or excessive. Where that power is lacking, the public services are starved. And the most amply provided of all public services are those where public bureaucracies are symbiotically associated with the most highly developed technostructures of the planning system. So, to the student of the present argument, complaint over the present distribution of public resources will have no element of surprise.

3

There is increasing complaint over the way income is distributed in the modern economy. As this book goes to press (1973), an election has just been fought with this as a central issue. Numerous wealthy idealists have been forced to re-examine their idealism lest it prove expensive.

The neoclassical system accepts that differences in ability, energy and diligence are differently rewarded. Income accrues to property, and the ownership of property is not imagined to be equally distributed over the population. In some parts of the economy there are monopolies with their special capacity for enriching the recipients of monopoly gains. But the neoclassical system assumes that resource movement will work generally toward lessening inequality. And it does not allow of enduring differences in income among workers in different parts of the economy. As put by one of the most distinguished exponents of neoclassical and antecedent ideas:

> Competition tends to eliminate differences in rates of wages for similar workers in different occupations and geographical locations, for the worker who is in the job where wages are low will move to the higher paying job. This movement will raise wages in the market the workers are leaving, and lower them in the market the workers are entering.
>
> Equilibrium will be reached in the occupational and geographical wage structure when the net advantages of all occupations open to the worker are equal. "Net advantages" embrace all the factors which attract or repel a worker, and the main content of the theory of competitive wage structure consists of the analysis of these factors.[1]

The present system leads, in contrast, to the expectation of

[1] George J. Stigler, *The Theory of Price* (New York: Macmillan, 1966), pp. 257–258.

an enduring difference in income between workers in different parts of the economy—and more generally. Five factors, all intrinsic to the system, suggest this result.

(1) The planning system resolves its conflict with unions by extensively conceding wage claims, including a share in productivity gains. The market system lacks both the power and, with some exceptions, the productivity gains.

(2) The control by the planning system over its prices and costs includes control over prices charged and paid to the market system. This allows it a substantial measure of control over its terms of trade with the market system. Being able to regulate its terms of trade, it naturally turns these (prices charged the market system, prices paid the market system) to its own advantage.

(3) This advantage is enhanced by the fact that the small entrepreneur in the market system remains in business in part because of his ability to reduce his own wage and in part because, unions being less frequent and less encouraged, he can on occasion reduce the wages of his workers. This exploitation has the sanction of the state, and the entrepreneur's self-exploitation is esteemed by the convenient social virtue.

(4) Measures to arrest inflation, as employed in the past, curtail demand, prices and incomes in the market system. In the planning system prices are under control, and wages are subject to the authority of strong unions. The effect of restriction in demand in this part of the economy is thus on output and employment. Incomes of employed workers, overtime pay apart, do not fall. And the unemployment sharply reduces what chance there may be for workers in the market system to move to the higher-paid employment of the planning system.

(5) The planning system, especially in the technostructure, requires personnel of relatively high educational qualification—engineers, other technicians, accountants, lawyers, statisticians, computer programmers and many others. Educational requirements for employment in the market system—most notably in agriculture—have traditionally been much lower. The quality

of the educational system accommodates to this difference. In addition some parts of the market system, agriculture again being the notable case, have relied extensively in the past on black and Mexican workers. Poor education, racial discrimination and the need to migrate from traditionally agricultural regions to areas of industrial employment have increased the wage differential between the market and the planning systems.

4

That there will be a continuing inequality as between the market and the planning systems (as well as within the planning system) is the clearest expectation from this analysis. It means that the broadly equalizing assumption of neoclassical economics must be rejected; instead, in the absence of energetic reform, the tendency of the economy is to one comparatively affluent, one comparatively impoverished working force. It is a conclusion that the circumstances of life in numerous urban ghettos, migrant camps and rural slums make real.

Indeed, to the intelligent lay reader, the need to prove a tendency to inequality of reward among different parts of the economic system will seem puzzling. Reference to low-wage industries is commonplace. The continuing poverty of numerous agricultural areas is notorious. So is that of the people who, having departed these areas, are still denied access to employment in the planning system and remain penned up in the urban ghettos. The statistics affirm both the difference and the unemployment. The answer is that the sedative formulations of the neoclassical system remain powerful. It continues to be taught and to be believed that, with economic growth and progress, inequality will decline. And, this being so, the moral energy of belief is only partially deployed on behalf of reform. Some obvious reforms—those that strengthen the bargaining position of workers and entrepreneurs in the market system in relation to the planning system—are made to seem unnecessary or unsound. Few differences can be greater than that between an economy in which such inequality is intrinsic and increasing and one in which it is exceptional and decreasing. And nothing is so important as instruction that identifies what exists.

5

It has come to be noted, more often in Europe than in the United States, that the people who have the most pleasant jobs —and who most strongly avow their devotion to their work— get the most money, those who have the least pleasant jobs get the least. The man on the assembly line or shop floor who would quickly disappear were his pay not forthcoming gets much less than the executive who tells, uncontrollably, of his joy in his job and the long hours he devotes thereto. The higher in the executive hierarchy, the greater, in general, the avowed enjoyment in work and the greater the pay. The absolute differences in reward between those who enjoy their work and those who do not is thus very great. Were we not totally accustomed to such an arrangement, it would seem remarkably bizarre.

Again this analysis provides the explanation. The planning system is highly organized; no convention is more firmly accepted than that one who is higher in an organization must get more than those below. Individuals naturally use their bureaucratic power to enhance this difference, and the higher one is in the corporate hierarchy the greater this power and this capacity. The result is a greatly attenuated salary pyramid with very high compensation at the top. This is explained, needless to say, not as the consequence of power but as the reward of the market for scarce and productive talent.[2] An agreeable formulation.

[2] The issue has been put cogently by Daniel Bell, "The Corporation and Society in the 1970's," *The Public Interest*, No. 24 (Summer 1971), p. 25:
Within the corporation itself, the differential between the lowest paid (often the common labor rate) and the average of the top executive group may be about 25:1 or higher. On what basis is this spread justified? The original rationale was the market. But increasingly the market becomes less relevant for the determination of the relative differences between "grades" of labor and persons . . . because human beings want and need a clear rationale for

The power of the planning system in the state ensures further that the income so distributed to the corporate hierarchy will be protected from adverse movement in taxation—and indeed will be favored. Virtually all recent changes in the tax laws have been benign in this respect down to such details as the classification of sybaritic consumption and erotic and alcoholic entertainment as essential to the conduct of business and thus a tax-deductible expense. The most spectacular success of the planning system, however, has been in having the income of the technostructure made subject to a maximum rate of 50 percent under the justification that, whatever the amount, it has the superior status of being earned. In the perhaps excessively restrained words of two of the nation's leading authorities on taxation, "The 50 percent maximum marginal tax rate on earned income . . . enacted in the Tax Reform Act of 1969 . . . was included in the act to provide tax relief for business executives and other earnings recipients who might otherwise be subject to marginal tax rates as high as 70 percent. Obviously, this provision—like many others—singles out recipients of certain types of income for preferential treatment while ignoring the fact that broadening the tax base would permit substantial reductions in all tax rates."[3] None of this, to the present reader, will seem abnormal.

6

The next complaint is that numerous products of modern industry have no serious function or do not safely or adequately perform their presumed function. And there is a growing suspicion of technological achievement in general. None of this the neoclassical system foretells. The failed invention is no extraordinary happening. Need is misperceived or, like the per-

the differences in reward among them, some principle of social justice for social distinctions will have to be articulated.
3 Joseph A. Pechman and Benjamin A. Okner, "Individual Income Tax Erosion By Income Classes," *The Economics of Federal Subsidy Programs,* A Compendium of Papers submitted to the Joint Economic Committee, Pt. I, *General Study Papers,* 92d Congress, 2d Session, 1972, p. 21.

petual motion machine, it does not work. But it is not suggested that people are served continuously with products that do no work or do not work.

In the planning system, we have seen, the test of innovation is not need but what can be sold—or what serves in the management of individual or public demand. In the case of consumers' goods a change that is without function may be as serviceable for selling a product as a change that has function —it may lend itself as well or better to commercials or salesmanship suggesting sexual fulfillment, sexual or social prowess, personal beauty, lessened obesity, personal or family prestige, preservation of youth or more efficient peristalsis. Or, being new, it may serve the simple supposition that what is new is better, with associated salability. This is normally more advantageous to the producer than the greater reliability that is associated with an older and proven product. Additionally, unuseful innovation will often advance the visual obsolescence in products. Complaint on all these matters and also on accidents of engineering that come with change for the sake of change is predictable in the present view of the economy.

Finally, in the case of public goods, weapons in particular, technological innovation (supported by the cooperative rivalry from other economic systems) serves to induce the obsolescence of the previous generation of weapons that leads, in turn, to their replacement. That this, in conjunction with the cost and lethal character of the products, should cast a dark shadow on the modern reputation of technological innovation is not surprising.

7

The next anxiety arising from the modern economy concerns its effect on the environment. The extent and depth of this concern need not be emphasized.

The neoclassical system recognized a flaw here. The price of a product or service might not include all of the costs of its production. Smoke, gas or smells might be passed into the air; wastes might be dumped into streams, lakes or the ocean; industrial or commercial development might assault the eye, although this last was not much mentioned. And it was the community, not the purchasers of the product, that paid. In some cases, as with increased outlays for soap or hospitals, a specific pecuniary cost was imposed on other people. Sometimes the general enjoyment of life was reduced. These were the external diseconomies of production—external because they were beyond the purview of the producing firm and not chargeable to it, and diseconomies because no one had thought to use the simpler term "costs." In principle there could also be external diseconomies of consumption—the costs imposed by the individual on others or the community from his consumption of a particular product, from the smog resulting from his automobile, the smoke from his tobacco, the aggression from his alcohol or the offal from his packaged food. However the external diseconomies of consumption were little discussed. Consumption was, as it remains in the neoclassical ideas, a nearly undiluted source of social satisfaction. Nor, in practice, were the external diseconomies of production taken much more seriously. They were an interesting theoretical imperfection in the market system, not a central aspect of its performance—". . . one of the chief obstacles to facile theorizing . . . rather than . . . an existing social menace."[4]

As regards even the market system neoclassical economics underemphasized external diseconomies. These could be considerable, as with pollution from cattle feedlots or the use of DDT and other pesticides on crops. Nothing is more destructive of the environment than the roadside commerce and unregulated urban sprawl of the market system.[5]

---

[4] Ezra J. Mishan, *The Costs of Economic Growth* (New York: Praeger, 1967), p. 56.

[5] The external diseconomies associated with modern agriculture rank in their magnitude with those associated with the production and consumption of industrial products. Feedlot fattening of livestock (which produce more organic waste than the total sewage from all municipalities in the United States), syn-

However, with the present view of matters, environmental anxiety becomes wholly predictable. The affirmative purpose of the planning system is growth. This becomes the purpose of the economic system and the society. The greater the growth, the greater, obviously, the impact on the environment—the greater the volume of waste to be disposed of in the air or water, the more countryside that is subsumed by industrial development, the greater the effect on the community of the resulting consumption. Also, since nothing is so important as an expanding output, there is a natural case for the highways, power lines, power plants, strip mines and urbanization that facilitate it. The claims of environment and amenity are inherently inferior to what serves economic expansion; they can be advanced only in the face of a heavy burden of proof.

We see also from this analysis the environmental consequences of unequal growth. These are profound. Economic development emphasizes the products of the planning system; it discriminates systemically against the civilian services of the state. In consequence it expands numerous types of private consumption with extensive external diseconomies—increased automobile use with its associated emissions and the spreading patina of abandoned and scrapped vehicle carcasses; increased use of packaged consumer goods with its associated litter of bottles, cans, cartons and nondegradable plastic; increased personal wealth with its increased rewards to larceny and violent assault and hence increasingly unsafe and unpleasant neighborhoods. And it accords no similar emphasis and support to the public services which make such increased consumption socially tolerable—which regulate automobile performance, provide substitute forms of transportation, pick up waste and suppress through police and courts the temptation to the direct appropriation of the increasingly promiscuous wealth.

We have seen also that the planning system has a high tech-

---

thetic nitrogen fertilizers and pesticides are particularly important. On this see Barry Commoner, *The Closing Circle* (New York: Knopf, 1971), p. 140 et seq.

nical dynamic. This means that it regularly substitutes new forms of pollution to which people are not accustomed for those to which they have become reconciled. It replaces fear of the known hazards of sulfur from burning coal with the unknown specter of radiation from an atomic power plant.[6]

Finally, when the public effect of the environmental damage becomes serious, the planning system (unlike the market system) has an alternative to remedial action. That is public persuasion. In the case of its products persuasion as to the reality is the substitute for the reality. So also with pollution. Instead of eliminating it, the natural recourse is to urge the public that it is imaginary or benign or is being eliminated by actions that are imaginary. In the first six months of 1970, firms in the planning system were estimated to have spent nearly a billion dollars to proclaim their concern over the problem of the environment.[7] One advertising agency promised for a mere $400,000, the cost of four two-minute television commercials for a total of 26 weeks, to do a major face-lifting service for any corporation that was under attack. Truth was not a barrier to such effort:

> . . . [the] "Chevron Research Center" . . . the impressive building in which Chevron . . . waged war against auto smog was the Palm Springs County Courthouse with a new sign . . . Potlatch Forests [said] "It cost us a bundle but the Clearwater River still runs clear." In response to criticism that the company had misrepresented its clean-up efforts . . . ex-President Benton Cancell said: "We tried our best. You just can't say anything right anymore—so to hell with it."[8]

The ideas here offered would appear to survive the test of anxiety over the environment.

[6] Ezra J. Mishan attributes special environmental effect to the kinds of industrial growth which are especially emphasized in the planning system, "in particular the growth of chemical products, plastics, automobiles, and air travel." "On Making the Future Safe for Mankind," *The Public Interest,* No. 24 (Summer 1971), p. 46.
[7] *Economic Priorities Report,* Vol. 2, No. 3 (September–October, 1971), p. 19.
[8] Ibid., p. 21.

8

There has been concern over the unresponsiveness of the modern large corporation to public will and its excessive power in the modern state. That the present ideas make this predictable need hardly be stressed.

The technostructure of the firm in the planning system pursues the goals that are important to itself—that serve its interests. Within an appreciable range of discretion it determines what it wishes to produce, establishes prices and persuades the consumer. And, in symbiotic relationship with the public bureaucracy, it similarly determines the product to be developed or produced and wins the assent of the legislature. That this should seem impersonal, arbitrary and bureaucratic to individuals who are taught that, whether in the market or at the ballot box, they are in command will not seem surprising.

The conflict is partly suppressed by persuasion. The larger strategy of salesmanship consists, in effect, in persuading the individual that the goals of the technostructure, including the pleasures that are associated with the purchase, use or possession of its goods, are the individual's own. But persuasion is by its nature imperfect. To those who are unpersuaded or insufficiently persuaded, the arbitrary nature of the system is fully seen or sensed. And it is possible that many who are conned have some inner sense of the wheel to which they are chained. In the case of public goods, weaponry in particular, persuasion is inadept or primitive. Often it is thought unnecessary. The armed services and the supplying industries are simply held to act out of a more intimate knowledge of need than the ordinary citizen can possess. Here, understandably, the sense of impersonal and arbitrary action is especially strong.

9

There is disparate growth as between industries with the result that some no longer supply what others require. In the case of gasoline for automobiles, fuel oil for furnaces, there is reference even to a crisis. Nothing of the kind is foretold by the neoclassical model. Given the model and its market, it could not happen. It is predictable when there is planning, and there is no reliable means by which the planning of different industries is made to jibe.

10

Finally, a matter treated in the last chapter, there is complaint that the modern economy shows a persistent and disconcerting tendency to inflation. And there is a growing disenchantment with the remedies people have long been taught to think right. That both inflation and its resistance to orthodox remedy are probable will not, in the present view of the economy, seem implausible.

11

A word is perhaps required on the source of the most urgent complaints. It is not the industrial labor force—the unions. This accords with expectation. The technostructure has come to terms with its labor force. Conflict is avoided by exporting the costs of settlement to other parts of the economy. The resolution is not perfect. But those who look for the classical class struggle within the modern planning system will look long and understand little.

Complaint, inchoate but increasing and on occasion violent, does come from the urban ghettos, from those who work for low wages in agriculture, from the young who have not found employment in the planning system. This too is predictable. *They* are the exploited. The present analysis does suggest that what is now attributed to race must be attributed also to such exploitation.

Strong complaint comes also from the university community. This too is predictable. The planning system requires a large supply of qualified people—its substitution in substantial measure of a lettered for an unlettered proletariat is one of its signal achievements. The emphasis of the older proletariat was on class discipline and class solidarity. The emphasis of the newer, educated proletariat reflects the values of the educational system that provides it. This emphasizes the worth of individual personality, commonly expressed as the importance of thinking for one's self. So, as the technostructure needs and seeks increasingly to persuade people, it also encounters more and more people, educated in accordance with its own requirements, who are taught to suspect such persuasion. Thus, in effect, the technostructure cultivates the criticism of its own need to override personality—to harness people to its purposes. This is a fact of first importance, a fulcrum on which much reform must rest.

12

That the economic system has a tendency to perfect itself will, perhaps, not now be believed. Unequal development, inequality, frivolous and erratic innovation, environmental assault, indifference to personality, power over the state, inflation, failure in inter-industry coordination are part of the system as they are part of the reality. Nor are these minor defects, in the manner of a misshapen wheel on a machine, which once identified and isolated can then be corrected. They are deeply systemic. They are part of a system in which power is exercised in unequal measure by producers; they derive from the exercise of such power. That power embraces, organically, the state—the normal source of reform. It depends, further, on its hold on our beliefs. And immensely relevant to what is believed is what is taught—essential to the power of the system is the view that most formal teaching provides of economic life. Anyone who speaks casually about reform in such a context is indeed being casual.

Yet the faults in the system are real—and painful and even oppressive. Imagery will work against everything but self-interest and the stubborn resistance of an always astonishing number of people to being fooled. Reform begins not with laws and the government. It begins with how we view the economic system—with belief. Belief may not be changed by argument. It is responsive to hard circumstance. For the pages that follow which deal with the general theory of reform, the preceding argument is vital. But the support of circumstance—the fact that things are not working—is even more important.

One other preliminary word: Talk of reform invariably arouses the indignation of those who associate wisdom with the belief that the condition of man is hopeless. The following pages will rightly arouse their rebuke. It may be justified. But their wisdom will indeed be demonstrated if we decide not to try.

FIVE

A GENERAL THEORY
OF REFORM

XXI

The Negative Strategy of Economic Reform

REFORM in modern economic society proceeds directly from the diagnosis. There would be little need for emphasizing such a notably unsubtle point were it not that the present diagnosis leads to remedies that are at variance, sometimes by 180 degrees, from the standard liberal or social democratic prescription. Yet, on reflection, this will not seem surprising. Were the standard remedies working, the anxieties just examined would not be present.

The invariable liberal or social democratic response to economic power is adverse. It is to dissipate, regulate, civilize or socialize it. In the United States the automatic liberal response on encountering industrial power is to call for vigorous action under the antitrust laws. Or the regulatory agencies—the Federal Trade Commission, Federal Communications Commission, SEC, FDA, Bureau of Standards, Department of Transportation or the Office of Consumer Affairs—are summoned to their duty. In its pricing, procurement, service or product design the corporation must be made to respect the public interest. Or, a very recent remedy, a campaign is mounted to place people who will speak for the public on boards of directors of offending corporations. Finally, if the complainant is radical, the offense extreme, the power of the offending enterprise great and the expectation of action minimal, there may be call for public ownership. Private enterprise has failed.

2

Mention has already been made of the useful futility of the antitrust laws. The practical experience cannot easily be ignored. The Sherman Act, the basic antitrust law, is only a little under a century old—it was enacted in 1890. The Federal Trade Commission, the principal agency of general industrial regulation, is about to celebrate its diamond jubilee. Were something to have happened, it would surely have happened by now. But we now see that there is more than long experience to induce doubt. The remedy implicit in the antitrust laws— the dissolution of the large corporate enterprise and therewith its power—is dramatic and even Draconian in aspect. Each new generation, as noted, can imagine that lack of past accomplishment was the result of the excessively pusillanimous tendency of their predecessors. All can assert hope, however exiguous, for the future.[1] Both the prosecution and the defense of the antitrust laws sustain among lawyers, in the manner of traditional automobile insurance, a rewarding pecuniary return. But from the standpoint of the technostructure and the planning system the antitrust laws are admirably innocuous. Were there only a handful of great corporations exercising power over prices, costs, the consumer and over public attitudes, perhaps their dissolution into smaller units and therewith the dissolution of their power might be possible. But a government cannot proclaim half of the economic system illegal; it certainly will not do so if its test of sound public policy is what, in general, serves the goals of this part of the economy. The planning sys-

[1] Not always, though, with great confidence. See Donald F. Turner's interesting essay, "The Scope of Antitrust and Other Economic Regulatory Policies," in *Industrial Organization and Economic Development*, edited by Jesse W. Markham and Gustav F. Papanek (Boston: Houghton Mifflin, 1970). Professor Turner, a firm supporter of the antitrust laws and a former Assistant Attorney General in charge of their prosecution, concedes "the past inadequacies and periodic atrophy of antitrust in the monopoly and oligopoly areas." But he contents himself with suggesting that "as a matter of public policy, there ought to be at least a modest expansion [in enforcement]." P. 76.

tem need fear only peripheral harassment by the antitrust laws. This has the principal consequence of persuading the public that something is happening.

But it will now also be evident that the antitrust laws, if they worked as their proponents hope, would only make problems worse. Their purpose is to stimulate competition, lower prices, otherwise unshackle resource use and promote a more vigorous expansion of the particular industry. But the problem of the modern economy is not the inferior performance of the planning system—of the monopolistic or oligopolistic sector, to revive the traditional terminology. The problem is the greater development here as compared with the market system. And the greater the power, the greater the development. Where the power is least—where economic organization conforms most closely to the goals envisaged by the antitrust laws—the development is least. If they fulfilled the hopes of their supporters and those they support, the antitrust laws would make development more unequal by stimulating development further in precisely those parts of the economy where it is now greatest. That would be unfortunate. If the expectations of the present analysis were realized—if the market were genuinely restored—competence would be lowered to the level of the market system. Policy unrelated to reality ends in absurdity.

Once it was argued that the antitrust laws, if vigorously enforced, would reduce the power of the monopolist to exploit his workers. Thus inequality would be reduced. Not even devout supporters of the antitrust laws now so argue, and we also have seen the reasons. The planning system does not exploit its own workers in the classical manner; in comparison with those in the market system, these workers are a privileged group. This is one reason why unions representing these workers do not share the liberal's passion for enforcement of the antitrust laws.

It will be evident that the antitrust laws are more than a blind alley along the path to reform. They are, as previously observed, a cul-de-sac in which reform can safely be contained.

3

The preceding chapters illuminate also the problem of regulation. If the purposes of the firm are believed to be those of
the public, there will be a heavy burden of proof on regulation
and the regulatory body. Unless there is a contrary showing,
it will be assumed that any particular intervention by the state
is against the public interest. In the name of the people it can
be righteously opposed. This is a severe handicap. And one
must expect that the planning system will at least partly capture
any regulatory body for its own purposes. Many have noted
that regulatory agencies tend to become the instruments, even
the puppets, of the industries they are supposed to regulate.
This we see to be normal.

Before the state can regulate in the planning system, it must
be perceived that public and planning purposes normally diverge and that realignment through regulation is natural, not
exceptional. And the state must be broken free from the power
of the planning system. To this we will presently return.

It will also be evident from this analysis that, some useful
offense to corporate dignity apart, nothing can be accomplished
by efforts to influence corporations by way of their stockholders
or boards of directors. The initial political task would be at-
tempted only by those who have not taken its measure. Voting
in the corporation is weighted pro rata with ownership. Such
is the distribution of share ownership that the votes of the few
and very rich invariably outweigh those of the many. What
is called corporate democracy may be compared, roughly, with
an election at large for the New York state legislature in which
the votes of the officers of the New York banks and the members of the Rockefeller family as a bloc are weighted equally
with and against those of the rest of the citizens of the state.
Not many legislators would be elected by the citizens.[2]

[2] The recent effort via the stockholders to force modest social reforms on
General Motors (Project GM), despite considerable effort and organization,

In the case of the corporation there is also the unfortunate absence of power in the event that a representative of the public *is* elected. The technostructure, we have seen, has the power which derives from knowledge and active participation in decision-making. With this, in the mature corporation, no board of directors meeting for a few hours monthly or quarterly can contend. A minority on a board of directors on which the majority is devoid of power can have little sense of its omnipotence. Such is the position of the public interest director.

Though innovative in other respects, the technostructure of the modern corporation is rarely adventuresome in political matters. Were it so, every large corporation would have on its board of directors a woman, a black, a devout ecologist, a consumer representative and the most ardent available exponent of safety. All known agitators would be so employed. All would meet every month or quarter with the board, ask searching questions, be advised of the value of their observations, be promised the most careful consideration. Nothing would happen. At least until the innocuousness of the arrangement was discovered, the planning system would be at peace.

4

The foregoing chapters also delineate the role of socialism as a remedy. The socialist, like the American liberal, is attracted to the positions of power. His antidote to the private exercise of power is public ownership. As with antitrust enforcement this does not remedy underdevelopment or exploitation of workers in the areas of power, for it is here that these ills are least. And public ownership is not a promising solution for

produced minuscule votes—never as much as 3 percent of the total vote on any issue. See Donald E. Schwartz, "Toward New Corporate Goals: Co-existence with Society," *The Georgetown Law Journal*, Vol. 60, No. 1 (October 1971), p. 57 et seq. General Motors did yield to the extent of appointing a distinguished black leader to its board. Professor Schwartz concludes, mainly from this, that "The results, apart from the voting, were very impressive." He is too easily pleased.

privately exercised power if the state itself is the instrument of such power. As with regulation the emancipation of the state from the control of the planning system must come first. Additionally the problem of power derives not from private organization but from organization. All organization excludes interference from outside or above; its goals are those which serve the interest of its members. This is the behavior of an organization before it is taken over by the state; it will be its behavior after it has been taken over. This will be especially certain if its operations are technical in character and its power is derived from more or less exclusively possessed information.

Goals may differ. A public organization will not need a minimum level of earnings to protect its autonomy. Technological virtuosity for its own sake may be more important than growth. But it will not be less concerned with pursuing goals important to its members than the private organization. Nor will there be any greater certainty that these goals will accord with public purpose. In recent times there has been at least as much complaint about the indifference of the Atomic Energy Commission to public interest as that of General Motors. In any superficial view the nuclear test explosion at Amchitka, even though ultimately harmless, was as impersonally indifferent to public opinion as GM is on automobile safety or exhaust emissions. Few will think the Department of Defense more subject to public pressure and concern than the American Telephone and Telegraph Company. This accords with expectations.

None of this, however, excludes a role for public ownership in the management of power when the latter is rightly perceived. Where public and private organizations react symbiotically with each other, there is power in the symbiosis—in the division of labor that allows of lobbying, deployment of political funds, encouragement of political action by unions and local authorities, access to the legislature, management of intelligence information by whichever organization, public or private, is best equipped for the task. Here, as we shall later see, there is a strong case for full public ownership. And the case for such public ownership becomes stronger as the state is broken free from the influence of the planning system. We shall also see that public ownership is indispensable—and almost certainly inevitable—in important parts of the market system where inability to deploy power and to command resources is the problem.

5

In considering the Soviet and Chinese economies all observers are struck by the emphasis accorded to indoctrination—to the massive effort, including appeal to Marxist-Leninist principles and, in the case of China, to the thoughts of Chairman Mao, to win consent for and commitment to the purposes of the state, which is to say the purposes of its planning. We have now seen that this tendency is not unique. Consent for the goals of the planners must be won by all planning systems. The techniques for winning employed by the planning systems of the nonsocialist states—the neoclassical ideas, the identification of the convenient social virtue, advertising and other direct propaganda, the enforcement by the Establishment of canons of reputable thought—are infinitely more diverse and subtle than those of the Communist countries. The ultimate need and purpose has much in common with theirs. Planning is planning, and acceptance of the goals of the planners by the public must be won.

The first step in reform, it follows, is to win emancipation of belief. Until this has happened, there is no chance for mobilizing the public on behalf of its own purposes in opposition to those of the technostructure and the planning system. The latter will continue to pursue its purposes under the protection of the belief that its goals are those that best serve the public. The unequal development, unequal income, unequal and bizarre distribution of public expenditures, environmental damage, discriminatory and ineffective stabilization policies will continue, for it will continue to be imagined that they reflect accidental or sui generis error. It not being believed that any systemic conflict between public purpose and that of the planning system is involved, no continuing effort will be made on behalf of the public.

More important still, the state will remain at the behest of the planning system. It is only with a public state (as opposed to one possessed by the planning system) that the foregoing reforms, and others, can be carried through.

The emancipation of the state begins—as the invaluable instinct for salvation has already indicated—with the legislature. This, not the executive branch of the government, is the natural voice of the public purpose against the technocratic purpose.

With the retrieval of the state for the public purpose, consideration of those reforms requiring action by the state becomes possible. These fall logically into three parts. There is first the need to enhance radically the power and competence of the market system—to enhance, affirmatively, its development in relation to that of the planning system and thus to reduce from this side the systemic inequality in development as between the two systems. This includes steps to reduce the inequality of return as between the planning and the market systems—to improve the bargaining power of the market system and reduce its exploitation by the planning system. This is here called The New Socialism. Necessity has already brought the new socialism farther into being than most suspect.

Then comes policy in relation to the planning system. This consists in disciplining its purposes—in making these serve, not define, the public interest. This means restricting resource use in the areas of overdevelopment, redirecting the resources of the state to serve not the planning system but the public, asserting the higher purposes of the environment, making technology serve public and not technocratic interest. These are the next steps to be considered in the strategy of reform.

Finally the economy must be managed. The problem is not to manage one economy but two—one that is subject to the market and one that is planned by its constituent firms. This management is the last step in adumbrating the general strategy of reform.

Let it not be imagined that on any of these matters action is novel. The problems to be solved are real. So are the inconvenience and suffering from leaving them unsolved. So in virtually every case practical necessity has already forced a measure of practical action in accordance with the expectations of the analysis here made. The present proposals have novelty, in most cases, only in being in conflict with the modalities of established neoclassical thought, in showing the theoretical justification for what circumstances and good sense have already initiated.

XXII

The Emancipation of Belief

THE EMANCIPATION OF BELIEF is the most formidable of the tasks of reform and the one on which all else depends. It is formidable because power that is based on belief is uniquely authoritarian; when fully effective, it excludes by its nature the thought that would weaken its grasp. It can also be pleasant —a womb in which the individual rests without the pain of mental activity or decision. Or, to change the metaphor, as with Tolstoy's happy soldier all personal responsibility is given over to the regiment. And the drums to which all march are those of others. The present task is to win freedom from the doctrines that, if accepted, put people in the service not of themselves but of the planning system. There is help from circumstances, for, as the previous chapters have shown, the practical consequences of this subordination—in uneven development, in unequal income, in proliferation of dangerous weapons, on the environment and on personality itself—are increasingly painful. Pain or even modest discomfort is better for persuasion than more abstract argument.

The belief to be contested is that the purposes of the planning system are those of the individual. The power of the planning system depends on instilling the belief that any public or private action that serves its purposes serves also the purposes of the public at large. This depends, in turn, on popular acceptance of the proposition that the production and consumption of goods, notably those provided by the planning system, are coordinate with happiness and virtuous behavior. Then all else becomes subordinate, more or less, to this end.

The virtue in question is that which is convenient to the purposes of the planning system. The virtuous head of a family works hard for an income that, however, is never quite sufficient for the things the family needs. These, as a practical matter, always increase a little more than income. If the man is on the shop floor, he is reliably on hand and welcomes any opportunity for overtime. He never stops working on the excuse that he has enough money for the moment and would prefer idleness. In recent years workers on the automobile assembly lines have taken to extending their weekends to increase their enjoyment of hunting, fishing, indolence and alcohol. None of these enjoyments is, in itself, thought wicked. But their pursuit in place of income is powerfully condemned.

If the individual is a professional or an executive, the foregoing compulsions are much increased. He is, in addition to all else, an example. So he is peculiarly relentless in his effort. He, of all people, cannot be negligent in his commitment to what is always called a better standard of living and on occasion is accorded the cachet of being called an American standard of living. The genuinely easygoing (as distinguished from the hard-driving) executive is a rarity. The bohemian executive—the corporation man who houses himself simply, dresses shabbily, feeds and entertains himself indifferently—is unthinkable.

The required virtue extends to the family. A wife is good if she devotes her time to procuring and processing the commodities and maintaining the artifacts that comprise the highest possible living standard. Sons are good if, whatever their adolescent oats, they settle down at the appropriate age to study engineering, science, business administration or other of the useful arts and become, with the aid of a like-minded woman, high achievers, the latter being the close synonym for those seeking high income and consumption. Likewise daughters if, after some juvenile experiment with art, indifferent dress or promiscuity, they adapt themselves to their mothers' mode of life. The virtue, it may be added, is heightened if the family accords support to measures—taxation, zoning, a reasonable attitude on air and water pollution—which ensure what is called a good climate for industry in the community, if they have an instinct for policies and politicians which promise a steadily expanding national economy and if they are patriotically sympathetic to the need for a strong national defense.

At all points the virtue of this family serves the purposes of the planning system. Even the process by which its virtue is kept convenient to the changing needs of the dominant economic power is impressively efficient. Once, when all possible income was wanted by the capitalist, it was the frugal workman, the poor but faithful bookkeeper, who invited praise and who was known to have a simple but rewarding life. Now, growth being a prime goal of the planning system, admiration is for those who spend and consume expansively and even accept indebtedness in order to maintain a decent standard of living.

It is possible to imagine a family which sets an income target as its goal; which has husband and wife share in the provision of that minimum; which makes a considered and deliberate choice between leisure or idleness and consumption; which specifically rejects consumption which, by its aggregate complexity, commits the woman to a crypto-servant role; which encourages self-fulfilling as against useful education for the offspring; which emphasizes communal as opposed to individual enjoyments with the result that it resists industrial or other economic encroachment on its living space; which, in its public outlook, sets slight store by increased production of the goods of which it has a sufficient supply; and which is indifferent to arguments for expenditures on behalf of national prestige or military power from which it derives no identifiable benefit. This family is not formally condemned as wicked. It is not ostracized by the community. But such esteem as it enjoys is the result, primarily, of its eccentricity.

2

The need, very simply, is recognition that our beliefs and the convenient social virtue are derived not from ourselves but from the planning system. This seen, we also see that the patterns of a successful life are many and that a successful economy may well be one that enhances the opportunity for exercise of such choice. The maximization of income and consumption is one such option. So is the maximization of leisure —so long, at least, as it can be done without imposing cost on others. So is the setting of a target income and consumption which, when achieved, allows of other and nonremunerative use of time. And there is, of course, an infinite range of choice in such use of time. And there are different patterns of life and leisure in different years or at different times in life.[1]

It goes without saying that any attack on existing belief and the associated virtue will not be welcomed by those who reflect the attitudes and needs of the planning system. Nor will it necessarily be welcomed by those who are chained by existing belief to the purposes of the planning system. All of us rejoice in the vanity that we are the agents of our own will. And, as just noted, the need to think deeply about life and to choose between alternative economic styles will be for many an onerous burden. Better the accepted patterns of life than the terrible costs of thought and choice.

It will also be urged that for many or perhaps most families alternatives to the existing patterns of life do not exist. The pressures of physical or otherwise inescapable need—for the minimally essential food, clothing, shelter, medicine, education—preempt all energy. For many in the market system this

[1] It has long been expected that young people will pursue income only to the extent required for consumption including such artifacts as a bicycle or an automobile or such services as travel. Then these are enjoyed. Only beyond a certain age does respectability require that the individual settle down, which means that consumption and income should be maximized as the planning system requires.

must be conceded. But it is the further truth that, as income increases, needs continue to preempt all energy and eliminate all choice. This is the artifice of the planning system. The belief that this is necessary is what is being here resisted. The prime purpose of improving income, and especially of improving distribution of income, should be to increase the number of people who are removed from the pressures of physical need or its equivalent and who are able, in consequence, to exercise choice as to their style of economic life.

3

The first step in the assault on belief is to show the source of the present myth. The next is to identify and render null the specific instruments for perpetuating the myth. These instruments are four, as follows:

(1) The present economic pedagogy. The role of this has been sufficiently seen. The service of present economic instruction is not to understanding but to the purposes of the planning system. It is designed, however innocently, to keep the individual from seeing how he is governed, to accommodate his views to the purposes of the planning system. The first step in breaking the thrall of present belief is to see the effects of the formal instruction by which it is perpetuated. This ought to be firmly in the minds of all who study economics.

(2) The present orientation of the educational system. Paralleling the explicit instruction of neoclassical economics are the implicit assumptions of the modern educational system. These, broadly speaking, make income and consumption coordinate with achievement. They hold scientific, engineering, business and legal instruction to be useful; instruction in the arts, and notably if it is creative in character, to be decorative or recreational. What serves the planning system is standard fare; the rest is justified negatively in academic oratory as something the civilized man does not neglect. Educators have a particular responsibility to see that education is not social

conditioning. This means the elimination of all distinction between useful and unuseful fields of learning, all suggestion that there is an economic standard of social achievement.

(3) The present reaction to overt persuasion. The present reaction to advertising and other forms of producer persuasion is, broadly speaking, one of resigned acceptance. Exaggeration is always assumed and mendacity normally suspected. But to yield is not thought damaging. The emancipation of belief requires presumption against all persuasion by the planning system. Such persuasion exists to impose the goals of the technostructure on the individual. The individual who wishes to be free cannot, therefore, accede. He must resist. Anyone who without thought yields to persuasion as to purchases, behavior or belief yields something of himself to the purposes of the planning system. Persistence in such refusal to yield has further favorable effects in breaking down the dependence of the media on the revenues from such persuasion, a matter to which I will advert presently.

(4) The manufacture of public policy. The nonelective voices of authority on public policy in the United States—on what is wise regulatory, tax, expenditure, military and foreign policy—are, in the main, three in number. These are the bureaucracy in the Federal Executive; the literate acolytes of the corporate community, most notably the lawyers; and the universities. Over all of these but especially the first two the influence of the planning system is strategic. The planning system is symbiotically related to the Executive branch of the government. The literate acolytes exist, in some degree, to articulate the needs of the planning system and with an aspect of impartiality that is not possible for members of the technostructure itself. Their power is given popular recognition in the everyday reference to the Establishment. The universities, as elsewhere noted, are a source of both dissident and supporting attitudes. Their emphasis on personality makes their members congenitally suspicious or skeptical of the purposes of the planning system. But the formal pedagogy, notably in

economics but also in political science, strongly upholds them.

Emancipation of belief requires that all policy emanating from the Executive and the acolytes of the corporate community be assumed, in the absence of contrary demonstration, to be what serves the needs of the planning system. No one should imagine that this service is conspiratorial, wicked or even deliberate. Much of the articulation will be by men of special unction and well-confined imagination to whom the notion of a divergence between the purposes of the planning system and those of the public is safely unrevealed.

More than policy is involved. It is important to recognize that as authority on public policy serves the planning system, so also does authority on public morality. Injunctions of public figures as to private behavior—as to work habits, hours, discipline, personal behavior as a consumer, what has recently been celebrated as the work ethic—will ordinarily be the voice of the planning system. Emancipation of belief requires that this also be known.

The political need for identification of the purposes of the planning system as something different from those of the public—what I call the Public Cognizance—is a matter to which I will return.

4

The power of the planning system rests on its access to belief. The question will be asked whether the emancipation of belief would not be advanced by a far more direct assault on the instruments by which it influences or controls belief. Why not outlaw neoclassical economics, ban advertising, break the hold of the planning system on the television networks or the other media, move in a major way to reallocate educational resources from the sciences and engineering to the arts?

There is no case for gradual means if more forthright ones will serve. But if belief is the source of power, the attack must be on belief. Law cannot anticipate understanding. And a deeper principle is involved. All planning seeks to win control of belief—to have people conform in thought and action to what, for the planners, is convenient. This is accomplished in the Soviet-style economies by formal promulgation—there is a correct and immutable *line* which, however, changes from time to time with the changing needs of the planners. In the Western-style economies the belief required by the planning system is achieved in the name of liberalism. It is held to be the set of conclusions to which men of good sense and reason come under scientifically acceptable instruction. Belief that is inconvenient to the planning system is not suppressed; it is either ignored or stigmatized as eccentric, unscientific, lacking in scholarly precision or repute or otherwise unworthy.

But because liberalism is a cover for convenient belief, liberalism is not wrong. The remedy is not illiberal suppression of the techniques for compelling belief but a truly liberal resistance to such belief. One does not suppress neoclassical economics; one shows its tendentious function and seeks to provide a substitute. One does not prohibit advertising; one resists its persuasion. One does not legislate against science or engineering; one sees their eminence in relation to the arts as the contrivance of the planning system.

Not only is this the liberal remedy; it is quite possibly the only one. Were suppression a practical or legitimate instrument of policy, it would be used not against the planning system but on its behalf. Nor is an emancipation of belief quite as remote and theoretical as it sounds. As people see that advertising, along with other commercial persuasion, has the purpose of subordinating them to the planning system—of putting them in the service of purposes that are not their own—there is a good chance that it will cease to be effective. As it ceases to be effective, it will cease to be used. Consumers will then have to pay for magazines, television and newspapers. And the media, in turn, will no longer be in the service, however subjectively, of the goals and values of the planning system. It is not beyond possibility that this development is already well advanced. Magazines that depend on advertising have been disappearing. The survivors are those that derive their revenue from readers.[2] Similarly, noncommercial and subscription cable television have been showing gains against that sponsored by advertisers. There is at least an instinct that the past emphasis on science and engineering had a tendentious industrial purpose. In recent years there has been an extensive revolt against the beliefs instilled by neoclassical economics. Finally there has been an encouraging increase in the skepticism with which the wisdom of the Establishment is regarded. In 1966, a poll conducted by Louis Harris showed that 55 percent of those sampled expressed "a great deal of confidence" in the nation's industrial leaders. By 1971, this was down to 27 percent. Confidence in bankers dropped from 67 to 36 percent; in educational leaders from 61 to 37 percent; in leading scientists from 56 to 32 percent; in advertising men from 21 to 13 percent.[3] Doubtless the disillusionment arising from the Vietnam war was a factor; confidence in military leaders dropped from 62 to 27 percent.[4] All these developments suggest, nonetheless, a healthy trend.

Finally in the last decade there has been a widespread rejection—notably among the young, at least while still young—of conventional standards of consumption. Dress, appearance, recreation, living arrangements and personal hygiene have all been affected. And with the rejection has gone a substantially diminished commitment to conventional career patterns. Back of this attitude, one senses, is an instinct that those involved in such careers are being used for purposes that are not their own. The change in attitude has occurred primarily among young people of middle and upper income families; it is in such families that the sterility of a life devoted to the competi-

[2] Another manifestation of doubt is reflected in the increased concern for truth in advertising and the increased demands on the Federal Trade Commission, the Food and Drug Administration and the other agencies responsible for enforcement of minimum standards of safety and veracity.
[3] Reported in *The Progressive* (December 1971), p. 13.
[4] Ibid.

tive pursuit of higher standards of consumption is most evident. Goods being of the least marginal urgency, their pursuit is enforced by attitudes of the most palpable artificiality.[5]

The revolt, significantly, developed during a period of comparatively high employment and of rather rapidly advancing income in the late sixties. Comparative ease of access to employment and income allowed, in turn, of part-time employment and led to a generally relaxed economic discipline. It was possible to participate in the revolt without suffering the sanction of total loss of income—the penalty classically imposed on those who do not accept the goals of the system.[6]

While the emancipation of belief has clearly begun, it has so far been subject to serious handicaps. The forces controlling belief have not been sufficiently identified. Nor the purposes of the control. Nor the techniques for such control. And, finally, it has not been seen that escape from the discipline of the planning system involves escape not from discipline alto-

[5] The existence of such a revolt was the thesis of Charles Reich's immensely popular *The Greening of America* (New York: Random House, 1970). There are faults in Professor Reich's book, including a considerable unclarity as to the nature and motivation of the economic system he foresees. But he rightly senses the discontent of younger people with a society in which effort is commanded by persuaded and competitive consumption. Both the audience won by the book and the peculiar bitterness with which it was attacked showed the sensitivity of the nerve that he touched. Clearly he was threatening the command mechanism of the established order. It may be noted, in this connection, that the standard (and highly effective) weapon of the defenders of established belief is to argue that any challenge is deficient in scholarship and therefore intellectually disreputable. This weapon was employed with vigor, although with only limited success, to my first venture with the present ideas in *The Affluent Society* (Boston: Houghton Mifflin, 1958). Its use is not always conscious. Scholars of the most mediocre mind are often the most vehement. That is because they are the most in the thrall of the convenient belief and hence the least capable of envisaging any alternative.

[6] Another factor which weakened the control of the planning system was the Vietnam war. Previously the rationale of the management of the public demand for military products was the need to match the military achievements of the Soviet Union or to serve the more abstract ends of national prestige and military security. These, if not universally accepted, aroused no overt hostility. And belief in the need for such expenditure did not overtly challenge intelligence. The Vietnam war aroused the immediate antipathy of those who might be called upon to serve in it. And the case for it was almost uniquely unbelievable.

gether but a transfer to the greater discipline of self. But, as the nature of the revolt clarifies, its spread is not in doubt. And there is precedent. In the latter decades of the last century and the early years of the present one, it was part of the public instinct that anything urged by the reputable community—by the bankers, the more respected newspapers, the more esteemed members of the United States Senate—should be regarded with suspicion. It reflected capitalist purpose. Its predictable effect would be the further enrichment of the already rich. Such was the deeper faith of populists in the United States and of socialists and social democrats in Britain and on the Continent. By the mid-thirties almost all reputable opinion was arrayed against Roosevelt and the New Deal. It was part of the popular instinct that this showed that Roosevelt was right. In the last thirty years the technostructure has accomplished the truly heroic task of recapturing public belief—of quelling this instinctive suspicion of the reputable view. Still it would be unduly pessimistic to imagine that the earlier capacity for skepticism has permanently been lost.

Also the human mind has a retrieving resistance to authority. When it perceives the processes by which, and for which, it is controlled, it is fairly certain to reject them.

XXIII

The Equitable Household and Beyond

> To effect her complete emancipation and make her the equal of
> men it is necessary for housework to be socialized and women to
> participate in common productive labour. Then women will
> occupy the same positions as men enjoy.
> —Lenin, *The Emancipation of Women*

THE MODERN ECONOMY, we have seen,[1] requires for its success a crypto-servant class. This class makes possible the more or less indefinite expansion of consumption in face of the considerable administrative tasks involved. One of the singular achievements of the planning system has been in winning acceptance by women of such a crypto-servant role—in making acquiescence a major manifestation of the convenient social virtue. And in excluding such labor from economic calculation and burying the separate personality of the woman in the concept of the household, where her sacrifice of individual choice goes unnoticed, neoclassical pedagogy has contributed competently to concealing this whole tendency even from the women concerned. It is now possible for most women to study economics without discovering in what manner they serve economic society—how they are being used.

However the acceptance by women of their major economic role is not completely secure. In recent years in all industrial countries there has been a measure of restlessness, even of minor revolt, among women. Again we have the validating reaction that shows the problem we are discussing to be real. As usual what is lacking is a clear view of the underlying

[1] In Chapter IV.

economic circumstance. And again emancipation of belief is itself a consequential reform. For once women perceive their role as instruments for expanding consumption on behalf of the planning system, their acquiescence in this role will surely diminish. Or so one may hope.

2

The reasonable goal of an economic system is one that allows all individuals to pursue socially benign personal goals regardless of sex. There should be no required or conditioned subordination of one sex to another. This, with the techniques by which such rights are presently denied to women, requires a substantial change in the way decisions on consumption are made within the household. These, at a minimum, ought to be appraised for their administrative cost to the participants. This means that the woman, as the administrator, should have the decisive voice on the style of life, for she shoulders the main burden. Or, if decisions are made jointly, the tasks of administration—cleaning, maintenance and repair of dwelling, artifacts, vehicles or planning and management of social manifestations—should also be equally shared. Either convention could, if adopted, bring a drastic change in present consumption patterns.

However the more plausible solution involves an attack on a more fundamental cause. That, at the deepest level, involves the concept of the family in which one partner provides the income and the other supervises the details of its use. To a substantial though unmeasurable extent the family is derived from economic need. For agriculture and handicraft production it was a highly convenient unit, one that involved firmly centralized responsibility for decisions on production and consumption and a useful division of labor on various tasks associated with both the production and use of goods. The man did the heavier field work or the more heroic tasks of the hunt or predation; the woman managed the poultry and made the clothes. With industrialization and urbanization men and women no longer share in tasks of production in accordance with strength and adaptability. The man disappears to the factory or office, the woman concentrates exclusively on managing consumption. This is a conventional arrangement, not an efficiently necessary division of labor; at a simple level of consumption it is perfectly possible for one person to do both. Without denying that the family retains other purposes, including those of love, sex and child-rearing, it is no longer an economic necessity. With higher living standards it becomes, increasingly, a facilitating instrument for increased consumption. The fact that, with industrialization and with higher living standards, family ties increasingly weaken strongly affirms the case.

It follows that, with economic development, women should be expected and encouraged to regard marriage not as a necessity but as a traditional subordination of personality, one that is sustained by custom and the needs of the planning system. It should be the choice of many to reject the conventional family in return for other arrangements of life better suited to individual personality.

This means also, as a practical matter, that women, if they are to be truly independent, must have access to income of their own. This is obviously necessary for survival outside the traditional family. And it makes possible an independent existence—for shorter or longer periods—within the context of the family.

With such income goes increased power over household decisions. With it also goes work which reflects, in some measure at least, the individual's preference as to how her days should be spent. Even if not satisfactory, as work for many is not, the choice is not preordained as is domesticity (meaning the administration of consumption) by marriage. Marriage should no longer be a comprehensive trap. A tolerant society should not think ill of a woman who finds contentment in sexual intercourse, child-bearing, child-rearing, physical adornment and administration of consumption. But it should certainly think ill of a society that offers no alternative—and which ascribes virtue to what is really the convenience of the producers of goods.

Equal access to jobs requires the support of law. It also requires a series of companion reforms. Not all of these are of great novelty, for the instinct for solving problems has, as usual, anticipated the clear diagnosis of the ill. Four things are of particular importance:

(1) Provision for professional care for children. The importance of this requires no comment. It is also, by the most orthodox measures of efficiency, a very great economic bargain. In a child care center one person cares professionally for numerous children; in the family one person cares unprofessionally for one or a very few. Thus there can be few institutions so directly designed to increase the productivity of labor. Professional child care has other advantages. The convenient social virtue ascribes to the present system the peculiar advantages of parental love. It almost never mentions the instances where this parental affection is diluted or negated by boredom, indifference, repellent personality, alcoholism, other psychological and physiological disorders or a grievous inability to understand those that beset a child.

(2) Greater individual choice in the work week and work year. That women need provision, as men do not, for maternity leave will be generally agreed. Their transition into the working force also requires flexibility. What could be accomplished gradually is excluded if abrupt reorganization of traditional household practices is thought to be required. A three- or four-day week, or a fifteen- or twenty-hour week, may make the transition possible for many women where immediate commitment to a normal working week would be an insuperable barrier.

There is a more general case for flexibility, one that is rewarding to both sexes. Only if an individual has a choice as to the length of his working week or year, along with the option of taking unpaid leave for longer periods, does he or she have an effective choice as between income and leisure. The absence of such flexibility—the commitment to a standard work week and year—assumes that all workers have roughly the same preference as between income and what it buys and leisure and the enjoyments that it allows. All, accordingly, conform to the same working week and year. This is a barbarous assumption, another denial of the individuality that the established economics pretends to defend. The desire for flexibility is evident in the widely differing preferences of workers as to overtime and moonlighting. In fact the continued commitment of industry to a uniform work week and year is a concession to the convenience of management and one that is by no means vital for efficiency.[2]

(3) An end to the present monopoly of the better jobs in the technostructure by males. For all practical purposes this monopoly is now complete; in 1969, in excess of 95 percent of all jobs in industry paying more than $15,000 a year were held by males. This reflects, in some degree, the sound instinct of the technostructure. If the latter is successfully to pursue the affirmative goal of growth, there must be women to administer the resulting increase in private consumption. Sensing this, the technostructure sets a sound example by excluding women from its membership, reserving them for the required household administration. It is in line therewith, as earlier observed, that the wife of the senior business executive is usually a model of acquiescent and prideful submission to competent household administration on a large scale. But one must be cautious in the search for a sophisticated explanation when a simple one is available. The male monopoly in business is for the beneficiaries an agreeable tradition; women have not chal-

[2] I have dealt with this in *The New Industrial State*, 2d ed., rev. (Boston: Houghton Mifflin, 1971), pp. 365–371.

lenged it because they are conditioned to the convenient social virtue—to the notion that family duty and the administration of consumption is their proper function.

The male monopoly will not be broken voluntarily. It will require the pressure of law, and since the corporation can no longer claim the immunity associated with assumed subordination to the market—since it must be recognized to be a public entity—there is no barrier to the application of such pressure. The appropriate procedure would be to require all firms in the planning system—all above a minimum size—to bring the representation of women at various salary levels into general accord with their representation in the working force as a whole. Ample time should be allowed for this; I have elsewhere urged, with colleagues, that major corporations be required to present a ten-year plan for bringing themselves into compliance.[3] No plea of intolerable damage to business efficiency can be allowed. On the contrary it will generally be agreed that women are not less stupid than men and that the intelligence available to the technostructure is, at any time, limited. Accordingly such a reform will bring a very great increase—potentially a doubling—of the supply of available intelligence.

(4) Provision of the requisite educational opportunity for women. The need for this is obvious. It is less obvious that, given the past discrimination, educational institutions, notably universities and centers of professional training, must for a period discriminate affirmatively in favor of women. To do otherwise is implicitly to perpetuate past discrimination. For the same reason the higher reaches of the technostructure must be under legal pressure to hire and advance women which, in

[3] Written in company with Professors Edwin Kuh and Lester C. Thurow of the Massachusetts Institute of Technology, and designated by us "The Minority Advancement Plan." See *The New York Times Magazine*, August 22, 1971. Similar provision is made therein for black and Spanish-speaking members of the community who, for other and even less defensible reasons, also suffer de facto exclusion.

turn, puts professional and other schools under pressure to provide them. When the effects of past discrimination have been erased—the same is true even more poignantly for the racial minorities that suffer discrimination—selection can become blind. But only then.

3

The consequence of the emancipation of women—and the rationalization of the household—will be a substantial change in patterns of life. Thus suburban life—as a wealth of commonplace comment affirms—is demanding in the administrative requirements of its consumption. Vehicle maintenance, upkeep of dwellings, movement of offspring, extirpation of crabgrass, therapy of pets, the heavy demands of social intercourse involving competitive display of housewifely competence are among the innumerable cases in point. Were all of these tasks shared by the male, there would be an urgent reconsideration of the advantages of suburban life. Urban multiple-dwelling units are much less demanding in their administration. Here, not surprisingly, wives of independent purpose in life are now generally to be found.

Lesser changes may be assumed. There will be more professionally prepared food. There will be less home cooking, the quality of which, though often dubious, is ardently celebrated in the convenient social virtue. Similarly there will be increased reliance on external services rather than home-installed machinery—on laundries, professional housecleaning and public transport instead of wife-operated and -maintained washing machines, housecleaning apparatus and automobiles. Professional entertainment will replace the social intercourse associated with exhibition of womanly talent in food preparation, home decoration, gardening and the dispensation of alcohol. Plausibly there will be an increased resort to the arts. The

arts, unlike, for example, competitive display of social crafts-
manship, are relatively undemanding in administration, and
they involve tasks which are themselves interesting and
preoccupying.

4

A final observation is in order. Economic production divides
between services and goods or things. With *many* exceptions
services are supplied by the market system; things come, in the
main, from the planning system. The rendering of services is
geographically dispersed and associated with the personality
of the individual performing them, both factors that lend themselves badly to organization and thus to the planning system.
Manufacturing of goods is pre-eminently a function of the
planning system—the great majority of the large corporations
are manufacturing enterprises. The consumption of goods, in
most cases, requires administration—preparation, cleaning,
maintenance, repair, disposal. The consumption of services
does not usually require administration; they are, in effect,
used up in the act of service. And a very large number of
services—those of laundries, garages, plumbers, snow removers
—have, as their purpose, the easing of the tasks of administration associated with the use or consumption of goods, including
real property.

It follows that, if women are no longer available for the
administration of consumption and the administrative task
must thus be minimized, there will be a substantial shift in the
economy from goods to services. This means, pari passu, a
shift in the economy from the planning system to the market
system. This, at least subjectively, could be sensed. It is, or
could be, another reason why producers of goods ascribe virtue
to the present role of women.

There are few matters on which the mind can dwell more
appreciatively than on the changes that would occur if women

were emancipated from their present service to the consumer society and the planning system. But it is the emancipation itself that is the present concern. Its further effects can now be left to others to envisage and to history to reveal.

XXIV

The Emancipation of the State

THE PLANNING SYSTEM pursues its own purposes and accommodates the public thereto. The government, through its procurement and in providing for the various needs of the planning system, plays a vital role in advancing the purposes of the planning system. Central to this function is the belief that what serves the purposes and needs of the planning system serves identically the public interest. What serves the planning system becomes sound public policy.

We have now a clearer view of the reality. In the planning system the purposes of the producer are dominant. The subordination of the state to individual purpose which makes producer and public interest identical is a disguising myth. Divergence between planning and public purposes, save by those who relish or are rewarded by illusion, must be assumed. Conflict is the general not the specific case. It will be convenient to have a phrase which bespeaks political recognition of this inherent conflict. As I have previously suggested, it may be called the Public Cognizance. The public cognizance accepts the basic divergence between the goals pursued by the planning system and what serves the public need and interest.

A cognizant public will expect the government, in the absence of countering effort, to support that part of the economy which is most highly developed. Thus government will add to inequality or imbalance in development. In doing this, and in responding otherwise to the needs of the planning system, it will contribute to the inequality of income distribution. Support to technical development will reflect the purposes of the planning system and, in the case of weapons, with a disenchanting potential for ending life. Public defense of the environment will be compromised by the higher purposes, notably growth and technical development, of the planning system. And the association and symbiosis between the planning system and the state will lead to a highly uneven development of the services of the state itself. Those services that reflect the needs of the planning system, procure its products and especially those that result from symbiotic association between technostructure and public bureaucracy will be amply provided. Those not so favored will be starved.

The role of the government, when one contemplates reform, is a dual one. The government is a major part of the problem; it is also central to the remedy. It is part of the problem of unequal development, inequality in income distribution, poor distribution of public resources, environmental damage and bogus or emasculatory regulation. And it is upon the government that reliance must be placed for solution.

Both roles of the government require the same remedy—that it be broken free from the control of the planning system. Until the government is so emancipated, simplistic proposals for government action will be useless. Nothing will happen or the effort will be turned, like so many others before, to the purposes of the planning system. No one can appeal with confidence for vital therapy to the local doctor if he serves even more devotedly as the local undertaker.

2

In the United States emancipation must now be recognized as the major issue—indeed *the* issue—in all electoral process. The word "recognized" must be emphasized; without being so identified, emancipation is already of the utmost importance. In presidential politics in the United States the Republican Party has come to be accepted as the instrument of the planning system. Its support from, and identification with, major corporate interest is not something which, even in the rhetoric of politics, it is any longer deemed possible or even necessary to deny. In the Democratic Party the position is much less candid. In an ideal world as viewed by the planning system, it would control presidential politics in both parties. This usefully reduces the risk of loss of control in any particular presidential election. Without wholly admitting its willingness in this regard, one part of the Democratic Party does accept the general principles of service to the planning system and the associated rewards. Support for weapons procurement, highway construction, technical support to the planning system on such projects as the SST, financial guarantees as in the case of the Lockheed Corporation, a passive acceptance of the tax privileges of the planning system and its participants—are all manifestations of this identification. So, of course, is the reliance of presidential candidates on the planning system and its participants for financial support. (A perfect record of upholding planning goals is not required. The planning system is tolerant of deviation on individual matters.) It is sensed, perhaps shrewdly, by this part of the Democratic Party that electoral success is only possible if the basic purposes of the planning system are accepted.

However in recent years another wing of the Party—though again without clear identification of its own character—has been associating itself with the public as opposed to the planning purposes. Basic needs of the planning system have been attacked. The election in 1972, with its emphasis on reduced defense expenditure, closing of tax loopholes, welfare reform, greater equality in income distribution and stronger measures to protect the environment and the consumer, involved an unprecedented attack on the purposes of the planning system. It also brought an unprecedented defeat. It is possible, however, that this defeat needs to be appraised in light of the newness of the effort. No candidate disposed to such general support of public purposes or such criticism of the purposes of the planning system had ever come close to nomination. It could have been a growing public cognizance that made this possible.

Needless to say, if there is to be any chance for emancipation of the state, there must be a political grouping that accepts the public cognizance and is expressly committed to the public purpose. In other industrial countries a similar political alignment must be expected, encouraged and welcomed. In both the United States and other industrial countries numerous more traditional liberal, labor, socialist and social democratic politicians, as they are variously denoted, whose habit of compromise with the purposes of the planning system cannot be broken, must be expected to align themselves with its defense.

The present political debate in the United States and other industrial countries is over varying interpretations of the public interest. The purposes of the planning system are held to reflect such interest. This is a highly misleading formulation, highly advantageous to the planning system. Given the public cognizance, the debate will be over the public interest as opposed to that of the planning system. These are the true terms of the discussion.

3

The litmus for the presidential candidate or his equivalent in any country is the public cognizance and a commitment to the public interest as distinct from that of the planning system. This is also the test for candidates for the legislature, and, in the strategy of emancipation of the state, the legislature must be regarded as playing a primary role.

This partly invokes the logic of exclusion. The public bureaucracy is the natural ally of the technostructure; in the developed case the two organizations are symbiotic. That a President must be tested by his grasp of the divergence between the public and the planning interests is obvious and urgent. But the President is, by the nature of his position, partly the captive of the bureaucracy; as Chief Executive he cannot be consistently at odds with the organization he heads. Such an adversary and disciplining role devolves, in part by default, on the Congress.

As repeatedly noted, all social reform projects a shadow; in this case it is marked. In recent years, on funds for an anti-ballistic missile system, a supersonic transport, space exploration, a manned bomber, the C-5A, the space shuttle, other military gadgetry, financial support to the Lockheed Corporation, the Congress has been the locus of opposition to the public bureaucracy where the latter was in symbiotic association with a corporate technostructure. All were cases of a marked divergence between the public purpose and the planning purpose—cases where the latter was served at particular cost to the former. This congressional reaction did not occur in a vacuum. It almost certainly reflected a growing public cognizance. Only lacking was a precise view of the forces giving rise to this reaction.

4

In past decades in the United States the Congress has been regarded peculiarly as the instrument of special economic interest. The Executive, in contrast, has been commonly thought of by liberals as the custodian of the larger public interest. The modern reversal of roles is not accidental; it reflects, like so much else in life, an accommodation to change in the underlying institutions.

In the early stage of corporate capitalism Congressmen and state legislators were the natural servants of the businessman in the affairs of government. The entrepreneur employed resources that were his own or which were extensively subject to his disposal. These—money and votes by employees along with the respect and fear inspired by the dependence of the community on his favor or on avoiding his retribution—were thus available to him as political power. Those elected behaved as befitted a subordinate being.

Additionally, in the case of the Federal Government, Congressmen and most Senators were elected from manageably small constituencies in which the money and influence of the dominant entrepreneurs could be brought effectively to bear. No legislator could be in doubt as to where support came from, what was expected in return or what would be the consequence of his failure to keep faith. However great his moral rectitude he adjusted his conviction, and therewith his conscience, to accord with political imperatives. The President, by contrast, had a much larger electorate. No firm or industry could claim appreciable credit for his election. He was also elected by a much more visible process which required that any surrender to economic interest be a good deal more discreet.

Among liberals the impression that the national Executive was above special economic interest was dramatically reinforced in 1933 and after, when Roosevelt made opposition to banks, big business and the rich the rhetorical (and, as regards prestige if not economic well-being, the real) foundation of his presidency. Having no other recourse, conservatives, mostly businessmen large and small, turned to Congress for their defense. Additionally the public bureaucracy that was created in these years was for the purpose of administering relief, social security, minimum wage legislation, support to unions, assistance to agriculture, rural electrification, development of the Tennessee Valley and support to public and private housing. None of these activities served directly the purposes of the emergent planning system; none had a symbiotic relationship with the large corporations. The new bureaucracies (and bureaucracy in general) were thus a natural object of attack by conservative and special economic interests in the Congress. The effect was further to strengthen the feeling that the Executive was the custodian of the public against the economic interest.

Since World War II, matters have much changed. As noted, services to the planning system have become far more important. The resulting bureaucracies—Defense, AEC, NASA, CIA

—are symbiotically related to the planning system. The Executive has come, in consequence, to be an expression of their interest, which is to say the interest of the planning system. At the same time the influence of the planning system on the legislature, though still powerful, has changed and possibly even weakened. The modern corporation does not have the same capacity to possess and control legislators as did the more elementary capitalist firm. The members of the techno-structure do not own the corporate financial resources they control. Thus they are somewhat more limited than the vintage capitalist in using these resources to purchase political support. Also the vintage capitalist was a local or regional figure; he could concentrate on the local or regional politicians whom he needed as the spokesmen for his interest. The characteristic corporation of the planning system is national in scope. There is some likelihood of disapprobation when a national corporation with headquarters in New York or Chicago involves itself financially or otherwise in a local congressional or senatorial contest. Better concentrate available resources and effort on the presidency.

Finally the planning system relies for its public power on the access to belief—on its ability to persuade legislators, as also the public, that its needs are identical with sound public policy. But members of a legislature are open to countering persuasion —and the people who elect them are subject to the pressure of circumstances hostile to the persuasion of the planning system. The latter suffer from the unequal development, unequal income, pollution or defects in product performance which result from the assumption that the planning interest is the public interest. So, as viewed by the planning system, the legislator is not reliable. There is always danger of his backsliding to the public interest. The original capitalist was more fortunate; for political reliability nothing approaches the man who is owned either by money or fear of certain defeat in the event of its loss.

5

In the neoclassical system, as in the popular myth, the voter instructs the legislator and the legislator relays the instruction to the public bureaucracy. The capitalist sought to alter this arrangement, not without success, by winning control of the legislator. The planning system also has a substantial measure of control over the legislature through its management of belief. But it has a much more forthright and direct control of legislators by way of the public bureaucracy.

This comes partly from the rewards that firms in the planning system and the symbiotic bureaucracies can distribute to compliant legislators. This control is reinforced in turn by the congressional committee structure and the associated seniority and tenure systems. To begin with the tenure system, there is a presumption that a member of Congress, having once been elected, should, in the absence of peculiarly extravagant personal or public dereliction, be re-elected. On being so re-elected, he gradually works his way up through the committee structure of the Congress without burden of displaying merit. This has the twin effect of removing him ever farther from the public pressures that reflect the public cognizance and of bringing him into ever more familiar relationship with the public bureaucracy with which his committees are associated. The most powerful committees—Appropriations, Armed Services, Ways and Means, Finance—are those associated with the largest and most powerful public bureaucracies. These are among the most desired. Here the needed seniority—the required divorce from the public cognizance—is greatest. Additionally legislators sympathetic to a particular bureaucracy tend to request service on the committees handling legislation or appropriations for that bureaucracy.

Finally there are the direct rewards to the compliant committee members or chairmen. Where the capitalist firm once paid its legislative servants with comparatively modest personal and campaign support, the Army, Navy and Air Force now give them expensive military installations, industrial plants and contracts. The Congressional District of the late Mendel Rivers, Chairman of the House Armed Services Committee, was so thickly covered with such largess that fear, no doubt exaggerated, was occasionally expressed as to the ability of the underlying geologic structure to withstand the strain. The state of Georgia was similarly rewarded during the tenure of the late Richard Russell on the Senate Armed Services Committee. An element of reproach always attached to the older forms of political bribery by the capitalist. This is not true of the modern arrangement. On the occasion of the first public display of the C-5A Transport in Marietta, Georgia, the President of the United States publicly accorded warm credit to the skill with which Senator Russell had exacted such reward from the public bureaucracy.

The congressional seniority and committee structure does not escape criticism. It is regularly attacked as inefficient, incompetent, sluggish in its response to public need and will and as the legitimizing instrument for gerontocracy as modified by senility. This criticism partly misses the point. Its most vital function is to advance the control over the Congress by the public bureaucracy and ultimately by the planning system.

6

The needed reforms follow as ever from the problem. In all congressional elections—in all elections for a legislature—there should be a presumption not in favor of re-election but against it. This is of the utmost importance. Only a vigorously manifested public cognizance by an incumbent should ever justify an exception. Such a presumption against re-election will greatly enhance the likelihood that legislators will reflect contemporary public attitudes. It will ensure, as a matter of course, the replacement of those who, under present arrangements, are co-opted, willingly or osmotically, by the planning system. It will prevent the present gradual subordination of the legislator to the public bureaucracy. It will mean a sacrifice of experience. But what is called an experienced legislator in all but the rarest cases is one who is learned in the needs of the public bureaucracy, and notably of those branches associated symbiotically with the planning system.

Congressional committees are inevitable. But a presumption against re-election would also mean a more rapid turnover in committee membership. This too would help ensure against a comfortable and stable subservience by the particular committee to the bureaucracy. Bribery by way of political favors in the legislator's district would cease to be worthwhile, for those so suborned would soon be gone.

A presumption against re-election should be accompanied by action to end the seniority system. Although enduring tenure is more damaging than the seniority system, the two go together; seniority and the associated influence, including access to the relevant rewards by the bureaucracy, are part of the normal case for re-election. And what is now called a "powerful chairman" is, with rare exceptions, a man who exercises power in the Congress on behalf of one of the more powerful public bureaucracies, the most notable case being the Armed Services. Election by the other committee members of his party instead of automatic advance with age would require that the chairman be responsive instead to his fellow legislators. Having to serve them, he could no longer as now so single-mindedly serve the bureaucracy and the planning system.

An effective President is one who leads and disciplines the bureaucracy to public goals as distinct from those of the planning system. A weak President is one who surrenders to the symbiotic goals of the planning system and the public bureaucracy. But the Congress is strategic for the emancipation of the state. It is not part of the public bureaucracy; it is meant to be responsive to public concerns. With a Congress so motivated, a President has a chance of identifying and pursuing the public interest. Even a weak President will be moved in this direction. Without congressional pressure and support almost any President will be hopelessly the victim of the public bureaucracy and the planning system.

7

Given the public cognizance and emancipation of the state (a massive but central assumption), seven lines of public action become possible—and necessary. In all cases the pressure of public need has already forced some action in conflict with the approved belief of the planning system. The seven requirements are as follows:

(1) Measures to equalize power within the economic system. No remedy is possible that does not enhance the power of the market system or reduce the power of the planning system or both. Equalization of performance and equalization of income both require equalization of power.

(2) Measures directly to equalize competence within the economic system. Of particular concern here are functions—provision of housing, surface transportation, health services, artistic and cultural services—which do not lend themselves to organization by the planning system and which are not rendered competently by the market system. This defines a major area of social action—or socialism.

(3) Measures directly to enhance the equality of return as between the market and the planning systems and within the planning system to offset and hopefully to overcome the inherent tendency to inequality.

(4) Measures to align the purposes of the planning system, as these affect the environment, with those of the public. This includes the regulation or prohibition of such effects of production and consumption—pollution of air and water, damage to landscape—which serve the purposes of the planning system but are in conflict with the purposes of the public.

(5) Measures to control public expenditures to ensure that

these serve public purposes as distinct from those of the planning system.

(6) Measures to eliminate the systemically deflationary and inflationary tendencies of the planning system. These must not, as in the past, be a source of added power for the planning system. They must not discriminate against the market system. And they must be consistent with greater equality in income distribution as between the two systems.

(7) Measures to ensure the inter-industry coordination of which the planning system is incapable.

The foregoing categories of public action are the subject matter of the remainder of this book.

XXV

Policy for the Market System

THE UNEQUAL DEVELOPMENT of the economy is the consequence of the power deployed differentially within the planning system and as compared with the market system. And the
inequality in income as between the two systems has a similar
source. The planning system, in the most general case, has
power over the prices at which it sells to the market system
and over the prices at which it buys from the market system.
The terms of trade have, in consequence, a reliable tendency
to be in its favor. Much has been made in modern times of the
tendency of the powerful industries of the developed countries
to exploit, through their control of the terms of trade, the
weaker economies of the Third World. More mention might
have been made of the ability of modern large-scale industry
to exploit within its own country the small enterprise to which
it is far more intimately juxtaposed and where the opportunities for exploitation, accordingly, are far greater. That a
considerable amount of this exploitation is by the small entrepreneur of himself or his immediate family or of employees
who do not enjoy the protection of either the law or a union
does not alter the fact. Though he is praised by the convenient
social virtue, such praise is not, generally speaking, an adequate alternative to income.

It will be evident that any fundamental correction of the
relationship between the planning and the market systems
must begin with the equalization of power between the two
parts of the economy. This is no academic matter. It involves
intensely practical questions of how prices, wages and incomes

are established in the two systems—matters where, once again, necessity has already led to action that is not only unblessed by the accepted economics but is in conflict with its lessons. Viewed as we come to see it here, this action—agricultural price-fixing, other support to small businessmen, support to collective bargaining, minimum wage legislation, the proposed provision of a guaranteed minimum income, international commodity arrangements, even some protective tariffs—is an intensely logical response to the bargaining weakness of the market system. We have not moved too rapidly and incautiously with such legislation, as the accepted economics strongly holds. It is not a response to special circumstances or special hardship or politics, which is the normal explanation or justification. Given the structure of the modern economy— given the two systems—such action is the logical response to need. We suffer because we have moved too slowly and too cautiously and with too great a sense of guilt with measures to equalize the power wielded by the two systems.

Indeed few features of the neoclassical economics arouse more admiration for its effect than the way it rationalizes and conceals the disadvantages of the weak. One theory of the firm applies for all. There is, accordingly, no basic presumption of difference in advantage between one group of firms and another. Some firms do have control over their prices. But this does not distinguish large firms from small ones; one can have small monopolies as well as large. Also monopoly control is basically for the purpose of enhancing profits. It does not accord special access to technology,[1] capital or the state. And,

[1] In the standard pedagogy an overture is occasionally made to the argument that the monopoly, by virtue of its resources and its ability to exploit and protect its gains from innovation, may be more progressive than the competitive firm. And it is commonly held that technology has "given rise to economies of mass production which only large producers can realize." That the modern corporation and the power that it exercises are part of a larger accommodation to (among other things) the requirements of modern technology has, of course, no standing. For a standard but relatively progressive statement of the orthodox pedagogy on this matter see Campbell R. McConnell, *Economics*, 5th ed. (New York: McGraw-Hill, 1972), from which the above quotation (pp. 405–406) is taken.

although for no clear theoretical reason, the neoclassical monopoly is almost invariably discussed as it affects the consumer. Almost no attention is given to its control over the costs of the weaker firms from which it buys or to its control over the prices at which it sells to other and weaker firms. Thus the problem of the terms of trade within the economy, as these favor some firms and are adverse to others, is almost totally out of view.

It follows that any effort by small firms to combine, to stabilize or to enhance their prices is not seen as a response to their weakness in face of the market power of others. It is an original interference with the market—a step toward monopoly. So, obviously, is any government action with this result. So all such action is disapproved. The disapproval persists even though larger firms, as the product of their greater size and greater size in their markets, have such power as a matter of course. Combination or concerted action is outlawed even though its purpose is to allow the smaller firms to deal more effectively with their larger industrial customers—to turn the terms of trade in their favor. There is even deemed to be a modestly artificial quality about public efforts to supply capital or technical assistance to small firms—capital and technical competence which large firms obtain as a natural concomitant of size and power. The small firm that is subordinate to the market is greatly cherished by neoclassical pedagogy. Economists abuse that which they love.

2

Yet, as noted, what is ignored in theory exists in fact. Again circumstance forces the action that theory deplores. For close to half a century commercial farmers in the United States have sought and secured minimum prices for their principal products.[2] Similar action has been taken in all other major in-

---

[2] There were isolated earlier steps. The first support price for tobacco was put into effect in the James River Colony within a decade or so after the arrival of the first Europeans.

dustrial countries. In the negotiations leading to the formation of the European Common Market by far the most difficult problem was the reconciliation of the different price levels established by public action for farm products in the several countries. In the United States the Capper-Volstead Act earlier in this century accorded farm cooperatives a limited exemption from the antitrust laws for efforts to stabilize markets and prices. Purchasing cooperatives have long sought to influence or control costs of electric power, irrigation water, fertilizer, petroleum and other major farm production requirements and to ensure supplies at these prices.

Similarly retailers have won support from the state for price maintenance legislation, i.e., legislation to limit price-cutting or discounting that they cannot control. And under the Robinson-Patman Act small retailers have won protection against price-cutting made possible by the superior bargaining power of large competitors. Countless other small businessmen —taxicab drivers and fleet owners, liquor dealers, gasoline service station owners, parking lot operators—have invoked the support of the state in one guise or another for the control of their prices.

The classical case of market weakness is the sale by the individual of his own labor. The trade unions, somewhat exceptionally, have acquired general respectability as a device for redressing the bargaining weakness of the worker and have received the support of the government in doing so. And where the trade unions have not been effective, minimum wage legislation has been used to redress the weakness of the individual seller in the labor market. Only where the position of the worker is very weak or exploitation is blessed by the convenient social virtue are organization and minimum wage legislation lacking.

However there is disadvantage in action that is in conflict with the approved ideas. Those whose bargaining position in the economy is the weakest are likely also to be politically weak. Such is the case of farm workers, artists, numerous small tradesmen and owners of service enterprises. And since effort to redress weakness is unblessed by the approved economics, their need can be righteously ignored. It is far more difficult to force reform in opposition to the accepted current of ideas than on the breast of that current.

3

The foregoing defines the nature and scope of efforts needed to remedy the weakness of the market system. It means that not reluctant, not merely passive, but strongly affirmative support must be accorded to its efforts to develop market power. In a rational and fair economy neglected and weak groups would be sought out and assisted in developing such power. There would be a general presumption not against but in favor of collective action by those who are numerous, small and weak. In specific terms this means the following:

(1) *General exemption for small businessmen from all prohibitions in the antitrust laws against combination to stabilize prices and output.* Any resulting abuse should be cured by regulation—not by efforts to restore competition. Further, where prices and incomes are notably low or erratic, there should be affirmative public support to organization to stabilize prices and production and to regulate entry into the business. Agreements to this end should be fully enforceable. And there should be particular encouragement to collective action limiting self-exploitation—limiting hours and conditions of work of self-employed entrepreneurs, notably in service industries, retailing and small manufacture. Such action, supported as necessary by legislation, should extend to family-run enterprises. Although the convenient social virtue greatly defends the right of a parent to work both himself and his children to exhaustion, this may as often be a competitive need as a matter of parental preference and conviction.

The goal should not be in doubt. Nor should there be any effort at semantic concealment. The goal is to stabilize income and strengthen the bargaining position of the market system by collective and publicly reinforced power over the main determinants of income. An explosion of indignation over this reform—over such guild socialism—is predictable. This should be attributed not to the adverse consequences of reform but to the hold that neoclassical attitudes have over even the most elaborate minds. For the consequence of what is here proposed is only to give the small businessman (and his workers) something of the security in prices and income, and therewith in investment and planning, that the large firm (and its employees) enjoys as a matter of course. It gives the service station operator, auto dealer, shoe manufacturer a modest fraction of the market power that General Motors, Ford or Exxon take for granted. It gives small clothing or apparel firms some of the power in dealing with Sears, Roebuck and Montgomery Ward that the latter have in dealing with them. The indignation, let it be clearly noted, is over giving the small firm the same power as the large, including the power to deal effectively with the large.

(2) *Direct government regulation of prices and production in the market system.* As in agriculture, self-organization to support prices, and thus to improve bargaining power and to raise and stabilize return, is often impractical in the market system. Any individual producer can get the advantages of the improved prices resulting from the cooperative action without accepting the restraints on output that such action normally requires. When enough discover this possibility, the cooperative action breaks down. In these circumstances government regulation of prices and production must be regarded as a wholly normal policy. Since technology and associated capital investment require stable prices, the frequent effect of such stabilization will be greater and more efficient production than if firms are left to the unpredictable fortunes of the market. This, defeating the expectations of the accepted economics, has been all but universally the effect of agricultural price supports.

(3) *Strong and effective encouragement to trade union organization in the market system.* Nothing in economic life is more certain than that the evolution of economic society places the employee of the small firm in peculiar need of the support of a union. It is this employee whose bargaining position is weakest; it is his employer who survives because of that weakness. Even the briefest glance at the position of the farm worker, casual laborer, employee of the small service establishment, domestic worker, as compared with the trade union member in the planning system, affirms the point. Yet in the United States the law specifically denies the support of the National Labor Relations Act to such workers. It is hard to imagine a more illogical, even barbarous form of discrimination.[3]

(4) *An extension and major increase in the minimum wage.* The market system, as it now operates, is a design for allowing the small entrepreneur to survive by lowering his income. The first remedial step is to ensure that this is not involuntary—to allow collective action by producers to protect themselves against this tendency. And the second remedial step is to encourage trade union organization so that these employers cannot, at will, reduce the wages of their workers. For good reason such employers have a tradition of resisting trade unions. And, being small and geographically dispersed, these industries are often inherently difficult to organize. So the next line of defense is the minimum wage.

[3] To which, however, there is now a response. Organization in agriculture, though much publicized by the efforts of Cesar Chavez in recent years, is still primitive. Elsewhere the market system, together with state and local government where somewhat similar factors are operative, is by far the greatest area of union growth. Between 1960 and 1971, the union membership of Service Employees, a union clearly within the market system, increased by 76 percent. Membership in the Retail Clerks (a union extensively though not exclusively in the market system) increased by 90 percent. (In the same period, membership in State, County and Municipal Employees increased 150 percent.) Unions in the planning system had a much lower rate of growth—19 percent for the Automobile Workers, 4 percent for the Steelworkers and 0.2 percent for the Machinists. From *U.S. News & World Report* (February 21, 1972). The 1960 figures are from the U.S. Department of Labor. The 1971 figures are from the unions.

The minimum wage must be much more aggressively used than in the past. The test of the appropriate level, a most important point, is not what allows the small firm in the market system to survive. That is because the market system survives in part because of its superior ability to depress wages. Accordingly a minimum wage that is consistent with survival in the market system is one that perpetuates inequality. The ultimate purpose of the minimum wage is to eliminate the wage differential as between the market and the planning systems. This means, in effect, forcing those who patronize the market system to pay the full price for the product—one that reflects an equality of wage return with the planning system—or to go without. They must no longer be the beneficiaries of the bargaining weakness of workers in the market system. It is integral to this reform that no firm should be so small, no industry so weak, that (subject always to suitable time allowance for change and adjustment) it should be exempt from the minimum wage.

(5) *A revised view of international commodity organization and a cautiously revised view of tariff protection in the market system.* The large corporation extends its operations across international frontiers. Thus it is effectively protected against the special hazards of international trade. In other countries it participates in the oligopolistic convention that excludes price competition; foreign firms entering its home market operate under similar constraints. If its cost disadvantage is too great, it can produce in the low-cost country under its own brand. Having such protection, it does not need the further assistance of the protective tariff. It can take a large-minded view of the social merit of free trade.

Not so the firm in the market system. Its only hope for enjoying the same protection for prices and income that is enjoyed as a matter of course by the transnational firm will be an international agreement stabilizing prices and production by official action or a tariff. International agreements have been negotiated and operated with varying success in the past for wheat, coffee, rubber and sugar. The case for them is very strong. Less formal arrangements to stabilize prices or place quotas on shipments as between the market systems of different countries can similarly be justified. They only do for producers in such industries what transnational corporations accomplish far more efficiently for the planning systems. The difference is, that being visible, they arouse indignation.

The case for the use of national tariffs to protect the bargaining position of a national market system is less strong. Often there is advantage in replacing domestic with foreign production. The tariff is peculiarly likely—as international agreement is not—to invite retaliatory action with adverse consequence. Still where tariff action is required as part of an effort to enhance the bargaining position of a domestic industry—as will normally be the case—such action cannot be excluded on doctrinal grounds. To do so is again to deny to the market system a protection that the planning system builds and enjoys as part of its basic structure.

(6) *A strong presumption in favor of government support to the educational, capital and technological needs of the market system.* The planning system, we have seen, can be viewed as an adaptation to the needs of modern technology. It has the ability to supply itself with the requisite capital. And it has a singular capacity to have its public needs, including those for qualified manpower, become public policy. The market system lacks such power. Actions by the state to provide the small firm with research and technical support, capital and qualified talent reflect not preferential but compensatory treatment. They are also essential elements of any policy designed to reduce the inherently unequal development as between the two systems. And once again we may remind ourselves that conventional attitudes depend not on objective circumstance but on the control of belief by the planning system. Government support to research and education in engi-

neering or the physical sciences seems wholly plausible. The need for such research and manpower is in the planning system. The arts, we have seen, are in the market system. Support to painting, sculpture, local theater or television on an infinitely more moderate scale is regarded with grave suspicion. The first is sanctioned by the needs of the planning system and reflects its approved belief. The second depends on the lesser esteem of the market system.

4

Economic development, we have sufficiently seen, depends on power to command technology and resources—to initiate or adopt innovation and support it with the requisite capital, manpower and services of the state. Action to stabilize prices and allow of planning will, on occasion, as in the case of agriculture, enhance development. Its longer-run effect, accordingly, will be greater productivity and production and lower prices.

This, however, cannot be counted on. The primary purpose of the measures here mentioned is to increase the bargaining power of the participants in the market system and therewith their income. These measures do not provide a comprehensive planning power. But the most general and urgent problem of the modern economy is not the production of goods but the distribution of its income. That, accordingly, is the first thing to be remedied. There will be occasion later for considering means for redressing the inferior performance of key industries in the market system.

The effect of action to raise and stabilize return to workers and entrepreneurs will, in most cases, be higher prices for the market system in relation to those of the planning system. Only thus can inequality of return be corrected.

The effect of higher prices (in the absence of other action) will be smaller purchases, smaller output and less employment in the market system than would otherwise be the case.[4] This must be accepted. The market system now serves as an employer of last resort. By lowering their wage those who cannot find employment under the much more favorable conditions of the planning system are able to find employment—or employ themselves—in this sector. The reforms proposed here reduce or eliminate this possibility. Devout exponents of the accepted view have long held that action to enhance and stabilize entrepreneurial or worker incomes rewards those who are already in business or who already have jobs at the expense of those who are kept out because they are no longer allowed to offer their services at a lower rate. The argument has merit.[5] It gains further strength if, for reasons of educational disqualification, location in an urban ghetto or rural slum or because of race, those concerned are not readily employable in the planning system.

The solution, however, is not to enforce a permanently lower reward on all who are in the market system. Rather it is to bring return in the market system more or less into line with that in the planning system and to provide an alternative income to those who, under these circumstances, are not employed or cannot be employed. Only thus can the market system be kept from being a residual employer at rates below those of the planning system. Only by such action can there be an approach to equality as between the two systems without the weight of reform—any move toward equality—falling on those who are left without any income at all. Thus we come to the last and most urgent in this series of reforms. *That is the*

[4] There may be no absolute reduction; the general growth of the economy could well prevent that. And other things do not remain equal. A trend in the economy from products to services is generally assumed. And since the latter are more often supplied by the market system, greater growth in this part of the economy could offset some or all of the effect of relatively higher prices.
[5] However, as offered in the past, the argument has often been without merit. The higher minimum wage, it was argued, would lead to a substitution of capital for labor in a working force that is more or less uniform in quality. The result would be unemployment. Usually overlooked was the likelihood under such circumstances that the growth of the economy would bring a more-than-offsetting demand for workers.

provision of a guaranteed or alternative income as a matter of
right to those who cannot find employment.

As usual, circumstances have forced if not the action at least
the debate. Access to the comparatively well paid employment
of the planning system is now denied to a very large part of the
labor force—the planning system is a club to which only a
minority of workers belong. And not all who have flocked to
the industrial centers seeking admission have been able (or in
some cases willing) to find employment in the market sector.
In consequence it is deplored but taken for granted that a
large number of people in the best of times will require income
from public sources—will be on welfare. As this is written,
debate is proceeding in the United States on steps to regularize
this income and provide it as a normal aspect of the economic
system. It has been proposed at very modest levels by a conservative administration. And more generous standards have
been defended by more liberal aspirants for public office and
assailed as portending mass idleness, moral putrescence and
public bankruptcy. The irrelevance or banality of the discussion will not deter anyone who has followed this analysis. It
mirrors a situation that is real.

The guiding principle for an alternative income will be evident from the antecedent analysis. The level should be modestly below what can be earned in the planning system. This
will then set the standard of compensation that is required in
the market system. Those who are self-employed need not
then reduce their income below this level in order to employ
themselves. The alternative income thus sets a limit to involuntary self-exploitation. And similarly it sets a limit below which
wages in the market system cannot be reduced. A migrant
family of five will not be reliably available for farm work at
$3000 a year if a minimum of $5000 is available from family
income.

It cannot be held against the use of an alternative income
that some recipients will not work. It is right, as all present proposals provide, that the individual who works should get more

income than the one who does not. When he takes a job, he should lose some but not all of his alternative income so that he will always be better off working than idle. Work remains an inescapable requirement of economic society. But it is a central purpose of the alternative income that an individual not be forced to reduce his income below some minimum in order to get that work.

Nor can it be held against the concept of an alternative income that some economic tasks will no longer be performed. Numerous ill-paid services of the more derogatory sort—the man who shines shoes in a hotel or airport or tenders a towel in a washroom—are now performed by people who have no alternative source of income or, at a minimum, are persuaded by the convenient social virtue that reputability requires them to take useless and demeaning jobs. Given an alternative source of income, some so employed would not work. The services they render would disappear. This should be viewed not as a loss but as a modest advance in the general state of civilization. Those services which society does not compensate decently are not sufficiently important so that their passing need be deplored.

So much then for reforms designed to strengthen the bargaining position of the market system, its claim on income and its prospect for reasonable equality with the planning system. The key pillars are: organization by small businessmen and the self-employed to allow of some approach to parity of bargaining with the planning system; a far more vigorous use of the minimum wage; and strong support to trade union organization in those areas where, in the past, it has been least encouraged and where it is most needed. But adding strongly to the bargaining power these steps provide is the institution of an alternative income at generous levels. This is central to all practical hopes for overcoming the inequality that is systemic as between the planning and the market systems.

XXVI

Equality Within the Planning System

THE MARKET in the neoclassical system is the instrument for distributing economic resources—manpower and capital—among various uses in accordance with the ultimate instruction of the consumer. And it is the instrument by which such resources are valued and compensated, this compensation being in general accordance with the value of their contribution (technically at the margin) to the productive tasks in which they participate.

Where manpower is concerned, some scarce and useful talent gets a very high return. That, however, is no major ground for indignation or complaint. Men of such talent get much because they contribute much. Others of lesser talent get more because it is their very good fortune to be directed by, or otherwise associated with, those whose superior talents earn so much. It would be hard for the well-rewarded man to imagine a more amiable doctrine. He does well by doing well; because he does well, others do better than they deserve.

The reality of the planning system, we have seen, is more commonplace. Reward depends not on the market but on determination by human agency. That is the nature of planning. In the mature corporation tradition plays a part in this determination—no point is better accepted than that the boss should receive appreciably more than his subordinate, however more intelligent, energetic or efficient the latter may be. If the corporate hierarchy is very deep, as in a large corporation, the difference in recompense between those at the top and those at the bottom must, in consequence, be very great.

Power also plays a prime role in determining compensation. As a man proceeds up in the corporate hierarchy, his power increases. This power includes, inevitably, the power to influence his own compensation or that of the managerial category to which he belongs. This is a simple, not very controversial and extremely important fact.

Additionally, and finally, the compensation at different levels in the corporate hierarchy of one corporation becomes the standard for others. Executive and specialized talent is extensively interchangeable. If the standard of compensation in a particular category is higher in another corporation, executives will take their promotion in that firm rather than in their own. This will justify increases in the first firm. None of this need be in consequence of any shortage of executive talent; the supply of talent or its quality would not be less at the lower level of compensation. To be an executive would still be far better than fitting bolts on the shop floor. What this arrangement does serve is the fiction that compensation is decided impersonally by outside forces.

The notion of an impersonally determined salary scale is extremely important to its beneficiaries. The head of General Motors or ITT receives some fifty times the compensation of the typical man on the assembly line or factory floor—a considerable differential. The man so rewarded might hesitate to argue, himself, that he was fifty times as valuable. Were it agreed that this pay was a manifestation of his own inspired generosity, there might be complaint. But as one accepts the succession of the seasons, the acts of God and the onset of old age, one accepts the dictates of competition.

The acceptance in this case is all but perfect. Even the radical does not attribute the compensation of the head of General Motors, General Electric or General Dynamics to the power, guile or rapacity of those involved. It is what happens with capitalism and the market. Those involved are the innocent victims of their own good fortune. Nor does the trade union leader protest. If the executive can get it, he deserves it. The function of the honest union leader is to get more for his men, not to worry about what others receive.

Once the planning system is stripped of its market disguise, the way it arranges its compensation becomes a question of much interest and the proper subject of public policy. The differential that is set between those who get the most and those who get the least requires justification. And the conclusion that these differentials reflect an egregious and indefensible inequality becomes inescapable. There is no evidence and no reason to suppose that the supply of executive talent requires the stimulation of the present prices. The number of able and eager candidates is consistently large. Those who get the largest pay have the most pleasant jobs. They are also the people whose performance depends least on pay—those who most pride themselves on their moral commitment to their work. In contrast those who do the most unpleasant and soul-destroying jobs get the least pay. And these are the people for whom pay is most important for extracting effort. A visitor from another planet, and certainly one from heaven, would be astounded by this arrangement and even more by its acceptance.

The remedies, in principle, are four. Not only the level of wages but the differentials in compensation *should* be an issue in collective bargaining. Greater equality within the planning system should be a goal of tax policy. It should also be a central objective of wage and price policy. And the ultimate structure of the firm should be one in which these differentials are greatly narrowed.

2

In classical collective bargaining the interest of the owner or capitalist was opposed by that of the worker. The struggle was over the division of the return between the two. What the owner paid his manager or agent—the instrument of his will—was a detail with which the union was not concerned.

With the rise of the technostructure such attitudes become

sadly obsolete. The technostructure is now the original source of power. Its reward depends on that power and not upon its contract with the capitalist. And the amount so distributed is no longer a detail. The cost of sustaining the technostructure is a substantial part of total revenues. Its share in relation to what is received by workers can no longer be dismissed. The union which does not seek for its members an appropriate portion of the aggregate compensation is no longer doing its job.

And aggressive concern for income differentials, including the nonpecuniary perquisites of position, is especially important for white-collar workers. Here a permanently subordinate caste, consisting extensively although not exclusively of women, is consigned to a role of advertised inferiority. Not only is it agreed that compensation for secretarial, literary, communication, computation and like tasks should be far below that for executive functions, but it is also taken for granted that the privileged ranks will remain closed to those who perform these functions. This is important for perpetuating the caste differentials. Secretarial and like staff would not continue in a professionally obeisant role if it were encouraged to think of itself as eligible for the job of its employer. And the executive would feel less secure in his superiority and altogether less happy were this so. Accordingly the executive ranks are recruited separately from those who are assumed to have some special talent or training, and this segregation is further reinforced by the convention, all but universally accepted, that the senior executives in a major corporation should consist almost exclusively of white males.

The pecuniary rewards of rank are, as noted, supplemented by nonpecuniary perquisites. These are often less important for their actual content than for the way they proclaim superior (or inferior) status. The superior castes are accorded ostentatiously spacious, appropriately furnished and well-lighted working quarters, have special dining and toilet facilities, are given latitude to adjust their hours of work to personal preference or idiosyncrasy and are expected to adopt a suitably solemn or imperious manner. The subordinate castes are assigned to crowded working space and communal toilets, have more commonplace food and feeding arrangements and are expected to be hygienic in habits, restrained in dress, punctual in hours and reverent in general demeanor.

In recent times there has been some indication that wage differentials and the associated badges of superiority and inferiority are an appropriate object of collective bargaining or other group action. In Scandinavia, Germany and England trade union action has focused modestly on the peculiarly disadvantaged position of those who do the most unpleasant work. And in the United States women have shown signs of being less acquiescent in their permanently inferior role than in the past. Where very large numbers of white-collar workers congregate, as in the head offices of insurance companies or major industrial corporations, there have been occasional examples of action to eliminate such ostentatious badges of subordination as inferior food, regulation of attire or a requirement of manifestly servile behavior in the presence of those of higher rank.

However the chances of aggressive trade union action to narrow pecuniary and other differentials remain, at best, highly problematical. Where blue-collar workers are involved, the tradition of noninterference in such matters is strong. And it is matched by the obvious sensitivity of the technostructure to anything that appears to impinge on its autonomy including its right to do well by itself. The broad tendency is for the union not to fight the technostructure but to become aligned with it.

The tendency of white-collar workers to identify themselves with the technostructure is, if anything, even stronger. Most such workers are persuaded that their personal well-being is best advanced not by organization and the assertion of collective power but by cultivating the good opinion of the superior castes with which they are associated. In consequence they accept their subordinate role and identify themselves closely with the purposes of the organization of which they are a part. Such ingratiation, renunciation and subordination of personality to organization are referred to as loyalty. Such loyalty, in turn, is accorded a very high standing in the convenient social virtue of the planning system—a virtue to which those affected subscribe. The practical consequence is that the inferior castes among the white-collar workers are generally resistant to trade union organization.

Nothing would do more for equality within the technostructure than the strong development of the white-collar union, always assuming that it sees its task as transferring to its members some of the compensation and amenity with which those who have more power now reward themselves. It is not, however, a bright hope.

3

For promoting equality a reasonably equal distribution of income is much superior to an unequal distribution which is then remedied by taxation. Once people have income, they have a not wholly surprising resistance to action, however righteously inspired, to remove it. And their ingenuity in defending possession is great. Nonetheless progressive taxation is indispensable in the civilizing effort to attain a greater measure of equality in the planning system.

Such taxation gains strong support from an unclouded view of the nature of the planning system. So long as incomes in the technostructure are imagined to be set by the market, they have a functional aspect. They are what must be paid to get the needed quality and quantity of effort and to ensure that the supply of such talent is kept coming. To pay what is so required and then remove any considerable part in taxes defeats the very purpose of the payment. And some moral objections can also be adduced. The individual's reward is determined by his diligence and intelligence and the contribution of these qualities to social product. The state should surely be cautious about removing what effort and ability have shown to be a man's proper portion.

In keeping with this doctrine and reflecting also the tendency of the government to accept as sound economics the preferences of the planning system, the tax laws deal very gently with executive income. A substantial part accrues in the form of tax-free consumption. This consumption—official entertainment, recreation, travel and gifts—is held to be essential to the performance of business function though it is recognized informally by all to be, in fact, privileged enjoyment. Another substantial part of executive income is normally arranged to accrue as capital gains with a maximum rate of taxation of 35 percent. And, as earlier noted, a recent major concession limits the rate of taxation of salary income to a maximum of 50 percent.

The case for such concessions evaporates when executive income is seen to be a function not of market valuation but of tradition, hierarchical position and bureaucratic power. Since it is these, not effort or intelligence, that are the determining factors in the reward, there is no danger that the input of energy or intelligence will be threatened by increased taxation. Nor will their supply be increased by lower taxation. The only consequence of such argument is to perpetuate or accentuate inequality. The market is irrelevant. But its myth survives as a cover for tax avoidance by those best able to pay taxes.

Accordingly the present analysis supports the most vigorous use of the progressive income tax as an instrument for promoting equality. And it refutes the case for special treatment of what, with much license as applied to higher salary levels, is called earned income. And it similarly denies the need for special pecuniary incentives to executive performance.

4

The third instrument for equalizing income within the planning system is government action. This becomes possible, and in some measure inevitable, with wage and price control. With the development of the planning system public action to stabilize wages and prices is unavoidable. This dispels finally the notion that the market is the ultimate arbiter of income; in effect it means official acceptance of the fact of planning. Once wages are the object of official policy, it is not easily argued that executive salaries should be immune—although, since what serves the interest of the technostructure becomes sound policy, this case will be made. And once wages and salaries are subject to public action, nothing excludes the next step which is to narrow by public action the differential as between those who do the unpleasant work and those who like their work. If greater equality in income distribution is an objective of public policy, greater equality must be an objective in wage and salary control. It must be a goal of the latter to narrow differentials which are a reflection not of function but of hierarchy, tradition and power. This, in turn, could require a similar narrowing of differentials in government, in universities and in the professions. It would, for many, be a heavy test of their commitment to greater equality.

The most forthright and effective way of enhancing equality within the firm would be to specify the maximum range between average and maximum compensation. If wages must be set by public action, public action to establish such differentials between pay of workers and executives is no less legitimate. As price and wage control policies develop, this should be the goal. However a more gradual policy would be better than none. The overall goal of wage and price control in the planning system should be to hold constant the general level of prices while allowing wages to advance in accordance with gains in productivity. When combined with the goal of greater equality this means that the allowable gains should be concentrated more or less exclusively on the lower-paid workers, including lower-paid workers in the white-collar occupations. These, over the years, would then have a steady gain in real income while the income of those in the higher salary brackets would, at worst, remain constant.

As usual where the analysis identifies a course of action we find the latter in at least incipient form. No formal theory justifies public interference with executive income. As earlier noted, the identification of the Republican Party in the United States with the interest of the planning system is something which no one now seriously disavows. Yet, in 1971, when President Nixon was forced to fix wages and prices, he extended the action, at least in principle, to include executive compensation. No one argued that executive pay was a significant source of inflationary pressure. But even a President whose identification with the corporate community was not concealed could not hold that the market had given way to planning in determining the pay of the worker but not the executive. If the government was interested in the compensation of the one, it had to be interested in the compensation of the other. It is a logical next step for it to be interested in the relation between the two.

5

In its mature form the corporation can be thought of as an instrument principally for perpetuating inequality. The stockholders, as we have seen, have no function. They do not contribute to capital or to management; they are the passive recipients of dividends and capital gains. As these increase from year to year, so, and effortlessly, do their income and wealth. And the tradition of privacy accords the technostructure autonomy in setting its own compensation and in continuing and enlarging the differentials.

A solution would be to convert the fully mature corporations —those that have completed the euthanasia of stockholder power—into fully public corporations. Assuming the undesirability of expropriation this would mean public purchase of the stock with fixed interest-bearing securities. This would perpetuate inequality, but it would no longer increase adventitiously with further increases in dividends and capital gains. In time, inheritance, inheritance taxes, philanthropy, profligacy, alimony and inflation would act to disperse this wealth. Meanwhile the allowable differentials in compensation could be established by the state in accordance with what is deemed necessary and just.

In principle there would be no effect on management from such a change. The stockholder disappears, but the stockholder was previously powerless. Men of talent, even at the lower levels of compensation, would prefer the executive offices to the shop floor. And, in fact, there are numerous such public corporations—Renault, Volkswagen in its best days, the Tennessee Valley Authority, many publicly owned utilities—which are indistinguishable in their operations from so-called private corporations. In any case we are dealing here with the part of the economy which is characterized by relative overdevelopment. In consequence the social claims of efficiency are secondary to those of equality.

It will be evident that this discussion has brought us into what, in modern social and economic thought, is virgin and unexplored territory where even the bold do not venture. That there is a form of organization beyond General Motors and General Electric is excluded even from speculative discourse. It would be in keeping with what we regularly praise as our commitment to the canons of free and inquiring thought were we to dwell on the possibilities for corporate evolution, especially as these affect the chance for greater equality. The likelihood of ensuing action is not great.

There are, however, other steps toward socialism which are more immediate and pressing—areas, indeed, where under pressure of necessity and cloaked by various forms of disguise all industrial countries have already taken long strides. To this extant socialism we now turn.

XXVII

The Socialist Imperative

AS PREVIOUSLY SUGGESTED, no design for social reform is so completely excluded from reputable discussion as socialism in the United States. Its disavowal by the major political parties is assumed. Even the most radical candidate for office, if serious, follows suit: "I am certainly not advocating socialism." Frequently he explains that the measures he proposes, by their very radicalism, are designed to save the country from socialism. Free enterprise must be protected from the ill will that results from its own deficiency, excess or aberration.

More than economics is involved. There is also an association between free enterprise and personal freedom. Those who decry socialism are defending more than their power, property and pecuniary return. As also with the barons at Runnymede personal interest is reinforced by higher moral purpose. So high is this moral purpose that men of exceptional zeal do not hesitate to urge that advocacy, even discussion, of socialism be outlawed in order that freedom may be preserved.

The position in other developed countries is different in form but not greatly in consequence.[1] In Western Europe and Japan socialism is an evocative, not an evil word. The American result has there been accomplished by divorcing the word from its established meaning and even more completely from any implication of practical action. An Englishman, Frenchman or German can be an ardent socialist. But however ardent he is also practical. So he will not seriously propose that banks,

---

[1] One not unimportant difference will be noted presently.

insurance companies, automobile plants, chemical works or, with exceptions, steel mills be taken into public ownership. And certainly if elected to public office he will not press legislation to such end. However much he favors such action in principle, he will not be for it in practice.

The reasons for placing socialism under such stern interdict will, by now, be comprehensible. Socialism is not something that commends itself to the technostructure; the latter, having won autonomy from the owners, does not invite subordination to the state. Its protective purposes argue strongly to the contrary. The technostructure, as an autonomous entity, enjoys freedom in shaping its own organization, in designing, pricing and selling its products, in bringing its persuasion and therewith its power to bear on the community and the state and in compensating and promoting its own members. Instinct warns that this autonomy would be threatened were the technostructure an arm of the state. Then decisions on where to locate a new plant, on what executives are paid, on standards of promotion would be public business. As such they would legitimately be the subject of public comment and review and perhaps of public action. Thus the desire to protect the present fiction, which is that these are matters of no legitimate public concern; they are as the market makes them—purely private business.

2

But it would be wrong to associate the decline of interest in socialism exclusively with the needs of the technostructure and the beliefs thus induced—important as this influence may be. Democratic socialism (and revolutionary socialism, for that matter) has for long been in accord with classical and neoclassical economics in identifying and locating the central fault in economic society. That exists where there is monopoly power. Where there is monopoly, the public is exploited by smaller output than would be possible, at higher prices than are necessary. Given the power of the employer in the labor market, workers are paid less than could be afforded and is their due. They too are exploited. As in neoclassical economics the most pejorative term is monopoly, so in socialism it is monopoly capitalism.

The reader will see why the old passion for socialism has disappeared—or survives only in oratory and nostalgia. The monopoly behavior which was its original raison d'être does not exist, even though a tradition in socialist criticism requires that any such suggestion be condemned as an exercise in capitalist apologetics. The central problem of the modern economy is unequal development. The least development is where there is the least monopoly and market power; the greatest development is where there is the most. The more highly developed the firm and technostructure, the stronger, in general, its commitment to growth. A firm that is cozening its customers in order to expand sales cannot at the same time be exploiting them in the manner of a classical monopoly. This the public knows or senses. Only if one is deeply educated can one overlook the reality and be guided by the doctrine. The doctrine guides the deeply committed socialist, in the curious companionship of the neoclassical economist, to the wrong part of the economy.

Workers have deserted socialism for the same reason that consumers have. Workers, we have seen, are exploited—or they exploit themselves. But the exploitation occurs in the market system. In the planning system workers are defended by unions and the state and favored by the market power of the employing corporation which allows it to pass the cost of wage settlements along to the public. Workers in this part of the economy are, relative to those in the market system, a favored caste. The socialist guides attention to workers who are employed in industries of great economic power. These are the industries—steel, automobiles, chemicals, oil—where power is used, in effect, to meet the major demands of workers. Like the public the workers do not march. The American trade unionist disavows socialism. His European counterpart hears it advocated, applauds, but wants no action.

A final factor which this analysis also illuminates has weakened the traditional appeal of socialism. The modern corporate enterprise, as we have sufficiently seen, is highly organized—highly bureaucratic. So is, or will be, the publicly owned firm. When the choice was between private monopoly power and public bureaucracy, the case for the latter could seem strong. The public bureaucracy might not be responsive, but it was not exploitative and thus malign. The choice between a private bureaucracy and a public bureaucracy is a good deal less clear. A very great difference in substance has been reduced, seemingly at least, to a much smaller difference in form. Added to this has been the discovery that the larger and more technical of the public bureaucracies—the Air Force, Navy, Atomic Energy Commission—have purposes of their own which can be quite as intransigently pursued as those of General Motors or Exxon. Private bureaucracies rule in their own interest. But so do public bureaucracies. Why exchange one bureaucracy for another? As elsewhere noted, the declining appeal of the Soviet Union to the modern radical is here also explained. Why exchange one bureaucratic society for another? China and Cuba seem much more appealing models. Some of their appeal, alas, is the result of their lower level of development with, and for this reason, less elaborate organization—a defect they are strenuously anxious to correct.

3

Yet for those who hold the fortress against inconvenient ideas life is never simple. The same circumstances which have been reducing the appeal of the traditional socialism in the positions of power have been making a new socialism urgent and even indispensable elsewhere in the economy. The word indispensable must be stressed. The older socialism allowed of ideology. There could be capitalism with its advantages and disadvantages; there could be public ownership of the means of production with its possibilities and disabilities. There could be a choice between the two. The choice turned on belief—on ideas. Thus it was ideological. The new socialism allows of no acceptable alternatives; it cannot be escaped except at the price of grave discomfort, considerable social disorder and, on occasion, lethal damage to health and well-being. The new socialism is not ideological; it is compelled by circumstance.

The compelling circumstance, as the reader will have suspected, is the retarded development of the market system. There are industries here which require technical competence, related organization and market power and related command over resource use if they are to render minimally adequate service. Being and remaining in the market system, these they do not have. So they stay in a limbo of nondevelopment or primitive development, and, as development goes forward elsewhere, their contrasting backwardness becomes increasingly dramatic.

Adding forcefully to the drama (and the distress of those who would resist all thought of socialism) is the fact that certain of the retarded industries are of peculiar importance not alone for comfort, well-being, tranquillity and happiness but also for continued existence. They provide shelter, health services and local transportation of people. Housing in a cold climate, medical attention when one is sick and the ability to reach one's place of employment are remarkably unfrivolous needs. One can readily detect the hand of a perverse Providence in the selection of the retarded industries. He is clearly bent on bothering the truly pious free-enterpriser.

The failure of these industries to pass into the planning system has diverse causes. Housing construction and medical services are geographically dispersed. As with all services this militates against the development of a comprehensive organization and specialization at any particular point. Such division of labor as may be possible is accomplished with manifest inefficiency. The time of carpenters, plumbers and electricians or, in the case of medical services, specialized surgical practitioners, physicians or technicians cannot be so scheduled as to avoid long intervals of ineffective use or idleness.

Unions have also played a retarding role. They are not uniquely strong in these industries. But employers have been uniquely weak as in the case of the construction industry, compliant as in the case of the transportation industry or members of the union itself as in the case of the American Medical Association. Thus the unions have had a free hand in regulating or prohibiting technical innovation or (as in the case of the AMA for a long time) organization that would have allowed of more effective economic development. Finally, in construction and transportation, public regulation, often inspired by employees or unions, has acted to inhibit technical innovation and the associated organization.

4

There is only one solution. These industries cannot function in the market system. They do not develop in the planning system. They are indispensable as people now view their need for means to move about and for protection from disease and the weather. With economic development the contrast between the houses in which the masses of people live, the medical and hospital services they can afford and the conveyances into which they are jammed and the other and more frivolous components of their living standard—automobiles, television, cosmetics, intoxicants—becomes first striking and then obscene.

The impact of unequal development in the case of health and medical services is especially bizarre. Virtually *all* of the increase in modern health hazards is the result of increased consumption. Obesity and associated disorders are the result of increased food consumption; cirrhosis and accidents are the result of increased alcohol consumption; lung cancer, heart disease, emphysema and numerous other disabilities are the result of increased tobacco consumption; accidents and resulting mortality and morbidity are caused by increased automobile use; hepatitis and numerous disabling assaults are often caused by increased drug consumption; nervous disorders and mental illness follow from efforts to increase income, observation of the greater success of others in increasing income, the fear of loss of income or the fear of the various foregoing physical consequences of high consumption. At the same time medical and hospital care is not part of the development which induces these disorders. It lags systemically behind—for a large part of the population, including many who are relatively affluent, its availability is uncertain and its cost alarming or prohibitive. Again the hand of a perverse Providence.

The only answer for these industries is full organization under public ownership. This is the new socialism which searches not for the positions of power in the economy but for the positions of weakness. And again we remark that most reliable of tendencies—and the best of tests of the validity of social diagnosis—which is that circumstance is forcing the pace. In all the developed countries governments have been forced to concern themselves extensively with housing, health and transportation. Everywhere they are already, in large measure, socialized. This is true in the United States as elsewhere. Local and commuter transportation has passed extensively into public ownership. So, with the arrival of Amtrak, has intercity rail travel. In the United States the old, who combine exceptional medical need with inferior ability to pay—for whom, in other words, the market operates with peculiar inadequacy—are provided with medical and hospital care. There is a bewildering variety of public medical assistance to other individuals and groups. In the housing industry there is an even more intricate complex of publicly sponsored construction, publicly aided construction, publicly financed construction, public subsidies to private occupancy and public control of rents. These functions, in turn, are divided among federal, state and local levels of government in such fashion that it is doubtful if any single official in any major American city knows all of the public sources of support to housing in his community.

This is, however, a highly unsatisfactory form of socialism. The term itself is scrupulously avoided.[2] And the resulting action is not undertaken affirmatively and proudly with the requisite means, the best available organization and with a view to the full accomplishment of the needed task. Rather it is viewed as exceptional and aberrant. It requires apology. The most desirable organization is never that which is best but that which seems least to interfere with private enterprise; the test of result is not the full accomplishment of the task but what is sufficient to get by.

Only as socialism is seen as a necessary and wholly *normal* feature of the system will this situation change. Then there will be public demand for high performance, and there will be public pride in the action. This is not vacuous and untested optimism; proof is to be found in Europe and Japan. There, as noted, the word socialism is evocative, not pejorative. And while socialists in other developed countries are attracted theologically by the positions of power, they are not repelled by the need for public action elsewhere in the economy. This means that they can act pridefully in the market system. This has produced a radically superior result in the areas of weakness where socialism is compelled. Although there is much variation as between countries, urban land has been taken extensively into public ownership; a large part of all urban housing has been built under full public auspices and continued with full public ownership and management. Similarly hospitals are full public enterprises; doctors and other medical attendants are well-paid employees of the state. And it is, of course, taken for granted that public corporations will run the railroads and urban transportation. The performance of all of these industries in Britain, Scandinavia, Germany and

[2] The term socialized medicine was, until recent times, highly pejorative. It has now, one judges, ceased to be so. Unsocialized medicine is for so many so unsatisfactory and expensive that alternatives can no longer be condemned by adverse terms. Socialism, too many people suspect, might be better.

Holland is categorically superior to that in the United States. In other countries—France, Italy, Japan, Switzerland—the enterprises that have been fully socialized, notably rail and urban transport, are superior. Only those that have not been socialized are deficient. The difference between Americans and Europeans is not that Americans have a peculiar ineptitude for operating public enterprises. The difference is that Americans have been guided by a doctrine that accords a second-rate and apologetic status to such effort.

5

In the past the case for public ownership was conceded where, because of the importance of the service, as in the case of education or the national defense, or the difficulty in pricing it to a particular user, as in the case of road building or street cleaning, it could not be left to the market. Or public ownership was pressed where, as in the case of public utilities, there was an inevitable monopoly and thus a danger of public exploitation. With the rise of the market and the planning systems, and the consequent inequality in development, the case for public ownership becomes much more general. It is not that the market, though generally satisfactory, fails in particular cases. It is rather that the market system is generally deficient in relation to the planning system. Accordingly there is a presumption in favor of public intervention anywhere in the market system.

The point applies particularly to the arts. Unlike poor development in housing, health or transportation poor development of the arts does not inflict physical discomfort. But these enjoyments are in the market system; without special public sponsorship and support there must be a presumption of underdevelopment. People are denied pleasure and happiness that, with relatively greater development of music, theater, painting, would be theirs. Given the power, including the persuasive power of the planning system on behalf of its products and its development in general, a society where there is no public intervention on behalf of the arts and humane studies will be grievously unbalanced. It will have great wealth. And as compared with earlier periods when patronage of the arts was more generous its artistic achievements will be much less.

In the last decade or so the notion that the arts require special support in the modern industrial society has achieved a measure of recognition. And limited, even primitive, steps have been taken by way of public development of facilities and public sponsorship of work. Instinct as to the public need has again run ahead of the underlying theory. The present analysis shows an extensive and expanding sponsorship and support of the arts to be not only a normal but an essential function of the modern state.

Public intervention on behalf of agriculture—socialization of agricultural technology, support to agricultural prices to encourage and protect investment, cooperative procurement of fertilizer, petroleum and equipment, cooperative or public supply of electricity, subsidy in support of new techniques—is also essential for a balanced development. In the absence of such social action the supply of food and fiber would be insufficient and unreliable, the cost (like that of housing and medical care) very high. Here, however, the instinct which leads to action in conflict with approved principle but in keeping with the realities of economic life has been very strongly manifested. And the approval of farmers, if not of economists, has been sufficiently strong so that these tasks have been undertaken not with apology but with pride. Largely in consequence of such public action agricultural development has, at least until now, been relatively satisfactory in the industrial countries. Had agriculture been free from such public interference—had orthodox principle been controlling—the performance would unquestionably have been deficient and perhaps by now dangerously so. And agriculture would now be exhibiting, in incipient fashion, the weaknesses elsewhere associated with the market system.

6

Circumstances, it is evident, are not kind to those who see themselves as the guardians of the market economy, the enemies of socialism. And because it is circumstance and not ideological preference that is forcing the pace, there is little that can be done about it. Not even the epithet "socialist" can rewardingly be hurled at the individual who merely describes what must be done. Such is the case with the socialism so far described.

But the story is not yet complete. The case for socialism is imperative in the weakest areas of the economy. It is also paradoxically compelling in the parts of exceptional strength. It is here the answer, or part of the answer, to the power of the planning system that derives from bureaucratic symbiosis.

Where the technostructure of the corporation is in peculiarly close relationship with the public bureaucracy, each, we have seen, draws power from its support by the other. The large weapons firms—Lockheed, General Dynamics, Grumman, the aerospace subsidiaries of Textron and Ling-Temco-Vought—propose to the Pentagon the weapons systems they would find it advantageous to develop and build. The Department of Defense proposes to them the systems the Services would like to have. The resulting decisions are then justified either by the need to keep up with the Soviets or the need to remain ahead of the Soviets.[3] One or the other of these justifications is bound to succeed. As previously noted, not even the most devout defender of orthodox models risks his reputation for minimal percipience by arguing that the resulting output is in response to public will as expressed through the Congress.

Two bureaucracies, one public and one nominally private, are stronger than one. The public bureaucracy, in citing the need for new weapons, can seem to be speaking out of a dis-

[3] For a brief period in the late sixties the Chinese were employed in this role. This usage appears to have been discontinued as unduly implausible.

interested concern for the public security. Its control over intelligence allows it, as necessary, to exploit public and congressional fears as to what the Soviets are doing or might be doing. Commonplace procedure requires that any proposed new weapon be preceded by a flood of alarming information on what the Russians are up to. The private bureaucracy has freedom and financial resources not available to the public bureaucracy for making strategic political contributions, for mobilizing union and community support, for lobbying, for advertising and for public and press relations.

The combined power of the two bureaucracies would be usefully reduced by converting the large specialized weapons firms into full public corporations along lines mentioned in the last chapter. The government would acquire their stock at recently prevailing stock market valuation. Thereafter the boards of directors and senior management would be appointed by the Federal Government. Salaries and other emolument would henceforth be regulated by the government in general relation to public levels; profits would accrue to the government; so also would losses as is now the case. Political activity, lobbying and community persuasion would be subject to such constraints as a public bureaucracy must abide.

This change is one of form rather than substance. For the large, specialized weapons firms the cloak of private enterprise is already perilously, and even indecently, thin. General Dynamics and Lockheed, the two largest specialized defense contractors, do virtually all of their business with the government. Their working capital is supplied, by means of progress payments on their contracts, by the government. A not inconsiderable portion of their plant and equipment[4] is owned by the government. Losses are absorbed by the government, and the firms are subject to financial rescue in the event

[4] Information on such ownership is in Hearing before the Subcommittee on Economy in Government of the Joint Economic Committee, 90th Congress, 2d Session, November 12, 1968, Pt. 1, p. 134. It was furnished (at my behest) by some of the companies with notable reluctance.

of misfortune. Their technostructures are the upward extension of the hierarchy of the public bureaucracy; generals, admirals, subordinate officers and civil servants, on completing their careers in the public bureaucracy, proceed automatically and at higher pay to the corporate bureaucracy. The corporate bureaucracy, in return, lends its personnel to the upper civilian levels of the Department of Defense. The large weapons firms are already socialized except in name; what is here proposed only affirms the reality. As a rough rule a corporation (or conglomerate subsidiary) doing more than half of its business with the government should be converted into a full public corporation as here proposed.[5]

For unduly weak industries and unduly strong ones—as a remedy for an area of gross underdevelopment and as a control on gross overdevelopment—the word socialism is one we can no longer suppress. The socialism already exists. Performance as well as candor would be served by admitting to the fact as well as to the need. And in doing so we would be showing that the planning system cannot always make disreputable that of which it disapproves.

[5] I've discussed this proposal at more length in "The Big Defense Firms are Really Public Firms and Should Be Nationalized," *The New York Times Magazine,* November 16, 1969.

XXVIII

The Environment

. . . if we are to preserve a habitable earth we must be willing to accept fewer foods and services, including less electricity.
—Charles F. Luce,
Chairman of the Board of
Consolidated Edison Company

ALTHOUGH THERE IS CONSIDERED to be professional damage in the suggestion, not all economic relationships are complex, and an example is the relationship between economic development and the environment. Economic growth is a central goal of the firm; from this it becomes a central goal of the society. Growth being a paramount purpose of the society, nothing, naturally enough, is allowed to stand in its way. That includes its effect, including its adverse effect, on environment—on air, water, the tranquillity of urban life, the beauty of the countryside.

Environmental damage follows from both the production and consumption of goods—from the effect of a power plant on the air and the resulting neon on the eyes, of a steel plant on the adjacent lake and the ensuing automobile on the lungs. The damage can also be unitary or collective. It can be from a single paper mill desolating the nostrils or a hundred smokers or automobile owners doing the same. The difference is of considerable practical importance: The paper mill cannot deny its responsibility; the automobile owner can regret the general consequences of automobile use but have no individual sense of responsibility, for what he adds to the total damage is immaterial.

Environmental damage is not peculiar to the planning system. Farm feedlots, farm fertilizer use and the food vendors, service stations, motels, repair shops, pet hospitals, all claiming attention on the approaches to every modern settlement, attest the capacity of the market system for both physical and visual pollution. This damage is also under the protection of the belief that nothing should interfere with economic growth.

2

A substantial part of the remedy for environmental damage consists, simply, in willingness to spend public money to clean up. There is no way of having clean streets except by hiring a sufficient work force and buying a sufficient amount of equipment to keep them clean. If water riparian to cities is to be pure, there must be adequate sewers and adequate treatment plants for the sewage. Much of the problem of the environment arises from underinvestment in elementary services and plant for keeping things clean or cleaning things up. It is, as the next chapter again emphasizes, the kind of expenditure against which the modern economy systemically discriminates.

Beyond the provision of such public funds there are three possible strategies for protection of the environment, two of which are irrelevant or impractical. The first solution comes from neoclassical economics, which views environmental damage as a defect in the market system. Wastes are dumped free in the air and water; the community, not the consumer of the eventual product or service, pays for the resulting clean-up. By internalizing such external diseconomies—by requiring the producer and thus his customer to pay the costs of preventing pollution or by taxing to pay for the eventual clean-up or to compensate for the eventual damage—the defect in the market is eliminated, and the problem is thereby solved.[1]

On frequent occasion the prevention of environmental damage does require the firm (and eventually its customers) to assume the resulting costs. Beyond this the economist's solution touches only a minor part of the problem—it depends all but exclusively for its standing not on its efficacy but on the piety with which its proponents find universal virtue in the market. It is rarely a remedy for the damage to the environment from consumption—there is no very good way of making those who smoke in public places pay for the discomfort of those who do not. In the end one prohibits smoking. To assess to airline passengers the discomfort of the noise to people below is equally hopeless. To assess to the passengers on an SST the damage to the upper atmosphere is not only hopeless but ridiculous. Nor can this remedy deal with the visual or other ecological consequences of production. There is no way of measuring and assessing the damage of a power line to virgin countryside or of a pipeline to the local wildlife. Neoclassical economics, as even its most prideful communicants would agree, did nothing to prepare people for the explosion of concern over the environment—something that might have been expected from a good and competent science. So economists would be wise to be restrained in recommending remedies that grow out of these ideas.

The second solution correctly identifies uninhibited growth as the problem and proposes limits to growth.[2] The defect in this solution is that damage to environment is a clear and present danger. Any alleviation from a reduction in growth is a matter of years and decades. The time frames, as grave

[1] For a detailed and sophisticated exposition of this remedy, see Robert M. Solow, "The Economist's Approach to Pollution Control," *Science*, Vol. 173, No. 3996 (August 6, 1971), p. 498.
[2] See most notably Donnella H. Meadows, Dennis L. Meadows, Jørgen Randers and William W. Behrens III, *The Limits to Growth:* A Report for The Club of Rome's Project on the Predicament of Mankind (New York: Universe Books, 1972).

scholars say, do not match. Additionally a reduction in growth only becomes a decent remedy as the distribution of income becomes more nearly equal. Otherwise people or groups are frozen in the consumption they now have. There is now some evidence that protection of the environment is a more appealing proposition to the affluent than to the poor.

The third solution is to continue economic growth but to specify by legislation the parameters within which it can occur. These parameters define the permissible damage of consumption and production to the environment. The setting of these boundaries becomes a major—in some ways *the* major—task of the modern legislature. On occasion it involves the prohibition of particular types of production or consumption —the public damage, as duly assessed by the legislature, is greater than the public enjoyment from the service or product.

Two further and far from surprising features of this remedy may be noted. It makes the state, not the market, the arbiter and protector of the public interest. This is hardly astonishing when it is remembered that the problem arises because the technostructure and its purposes have already replaced the market and the purposes of the public. Public guidance and planning of production is replacing private guidance and planning of production. The second and marked characteristic of this remedy is that it is what is already being done. It is the remedy which circumstance and the absence of a more acceptable alternative are forcing upon the modern state.

3

In the past, when the divergence of some private or planning purpose from the public interest (with environmental or other effect) became intolerable, it was the practice to specify the broad legislative purpose and to pass enabling legislation. Then a regulatory agency, new or old, was given the task of framing the specific regulations that reflected legislative intent, including specification of the time period within which conforming behavior would be required. And the regulatory body had considerable discretion in enforcement. This greatly simplified the legislative task and allowed of what is called flexibility in enforcement. It has also been a generally admirable arrangement for those who do not wish to conform. It has allowed the planning system to bring its natural power in relation to the public bureaucracy to bear in order to minimize, postpone or negate action. Where this power is great—as in the case of the automobile, oil, pharmaceutical, chemical or like industries—the resulting regulatory effect has been extensively neutralized. This tendency—the vulnerability of the regulatory agency to such capture—is well recognized.

The more difficult, more time-consuming but far more effective course is for the legislature itself, as the primary custodian of the public interest, to specify the required result. Such action is condemned as inflexible—as putting business in a straitjacket. That should deter no one. A straitjacket is a reasonably accurate figure of speech for what is required.

Effective environmental protection requires explicit and unyielding legal specifications. The quid pro quo for the firm then is full autonomy within this framework. Once the limits on action are established, the technostructure should not be second-guessed as to individual decisions. The strategy of environmental protection excludes, by law, action that is inconsistent with the public interest but allows the firm the maximum freedom of decision as to how the conforming results are obtained.

As to production effects this means comprehensive specification of the wastes or heat that may (or may not) be exuded into air and water but minimal interference with how this result is obtained. And, a highly important matter, it means a rigid specification of what may (or may not) be done to alter the landscape. Visual pollution is not less important than that which results from chemicals. Only the low order of importance that the modern economy assigns to the arts and aesthetic considerations has made it so. But, again, once the rules are established—development within specified industrial areas, no transmission lines over the countryside—there should be freedom of decision within the rules.

For consumption effects the broad rule is the same. The amount of a noxious gas that an automobile can emit, the survival value of a detergent or container are specified, and the manufacturer is then accorded discretion as to how this result is achieved. However the control of adverse consumption effects frequently requires specification of the conditions for the use of products, e.g., tobacco consumption in public places—and increasingly it involves the partial or total exclusion of certain kinds of consumption.

Automobile use in the central city, aircraft travel adjacent to populated areas, supersonic travel, random residential use of land are all cases where the advantage to the particular consumer is outweighed by the adverse effect on the community as a whole. In the past the presumption has favored individual convenience even in face of larger social cost, and this has reflected the purposes of the planning system. The rational legislative decision requires the exclusion from consumption of products, services and technology where the social cost and discomfort are deemed to outweigh the individual advantages.

4

The theory of action just adumbrated not only imposes major responsibilities on the modern legislature but will seem a remarkably comprehensive invasion of what, in the past, have been described as the prerogatives of private enterprise. Four points bearing upon this are worth noting:

(1) That we shy away from such interference is the consequence of the effective conditioning to which we have been subject by the planning system and, on its behalf, by the established economics. This has created the presumption that the goals of private enterprise are coordinate with those of the public. Given the public cognizance—given recognition of

316   *A General Theory of Reform*

a natural divergence between planning and public goals—measures to effect alignment are not remarkable but normal. If there is to be private planning, it can only be reconciled with public purpose by public planning.

(2) The legislative strategy just outlined has already been extensively imposed by need. Specification of what can be done to air and water is already a major legislative preoccupation in the United States; so is the protection of the landscape, although still preferably in places as distant as Alaska. The production of numerous environmentally damaging products—DDT, mercury compounds, nondegradable detergents—has been prohibited or restricted. Consumption, as in the case of disposable containers, automobile use in large cities, air travel over metropolitan areas, is increasingly being restricted. And there is an unmistakable tendency for the Congress, reflecting the public cognizance and reacting to the vulnerability of the public bureaucracy to the planning system, to specify the requisite behavior by law. The supersonic transport was stopped by the legislature over the powerful opposition of the Department of Transportation, working in symbiotic association with the relevant firm in the planning system, namely the Boeing Corporation. As ever we find circumstances verifying the relevant theory.

(3) In recent times reaction to the effects of industrial development on the environment has caused deep and general suspicion of all economic growth. As noted, advocates of zero growth have appeared and become vocal. A case for diminished, zero or negative population increase is also based on environmental effects although without resolving the question as to what is improved if a somewhat smaller number of people consume twice as many goods. The present strategy of environmental protection does not exclude growth. It accepts the commitment of the planning system to its own expansion, as also its need for autonomy of decision. It undertakes to discipline that growth, align it with public purposes and do this under public auspices. Although the effect, it should be

assumed, will be a lesser rate of growth, it might wisely be regarded by the planning system as the most conservative way of coming to terms with the environmentalists and the concerns they reflect.

(4) In the recent past the defenders of the environment have, on occasion, shown a tendency to absolutism. Reacting to the past weakness of their position and to the urgency of their task, they have opposed any economic development that has obvious environmental effect. No oil refineries, no power plants, no bridges, no high-rise housing complexes. There is grave danger in this. While the environmental effect is adverse, the ultimate consumption is well-regarded, on occasion even by the defenders of the environment. Should it become possible to attribute shortages of heating oil, housing, even air conditioning, to excessive environmental zeal, effort on behalf of the environment could be seriously undermined. As with most other things in life, protection of the environment has a cost. The gains from the protection must be weighed against the costs.

5

While the legislature is the indispensable instrument of the public cognizance in general, and the environmental parameters in particular, it is an instrument of notable bluntness. It acquires its position not because of its excellence for the task but because both the market and the public bureaucracy are, or tend to become, instruments of the planning system. Still the task is not as difficult as might be supposed. The conditions requiring remedy have also an obtrusive bluntness. Where the conflict between private and public goals is important, it will be brought forcibly to the attention of legislators. This will even be true of the more arcane damage from projected technological innovation where, increasingly, the scientific community can be counted upon for warning. The writing of the appropriate legislation does require informed care. Information to this end must be organized. Decisions must also be taken as to priority in action—between what is important and what is unimportant. Watch must also be kept on the enforcement activities of the Executive as well as on its performance of the residual functions that, however specific the legislation, must still be left to the regulatory agencies. The protection of the environment has become a major legislative task. For this the Congress—and lesser legislative bodies as well—do not now have the required technical staff. Planning in this field having been made inescapable, this shortcoming must be remedied. Any significant legislative body should have an effective environmental planning staff.[3] And its right to all needed information must be strongly affirmed. Here, in another manifestation of that matrix which characterizes all social phenomena, we are aided by the public character of the planning system. The right to privacy, including the right to privacy on technical matters, is a prerogative of the market-controlled firm. It evaporates when the firm exercises public power. So, in keeping with the need to establish legislative parameters for the planning system, goes the right to the information that this requires.

[3] In the case of the Congress this staff should not be confined in its service to any one committee. While some have a direct responsibility for legislation or appropriations bearing on pollution, virtually every congressional committee from Armed Services to the District of Columbia takes action bearing on the environment. The need is for a strong technical staff that is available to all.

XXIX

The Public State

THE GOVERNMENT, we have seen, contributes notably to inequality in development. Where the industry is powerful, government responds strongly to its needs. And also to its products. It gives the automobile industry roads for its cars, the weapons industry orders for its weapons, other industries support for research and development. By the same token it is sparing in its support to the weaker parts of the planning system, more so to the market system, most of all to public need unrelated to economic interest. What is done for Lockheed is sound public policy. The purchase of pictures for the National Gallery makes dubious economic sense. Farm subsidies are notably more wasteful than those for airlines. As noted earlier, it is no longer thought necessary to conceal the commitment of the Republican Party to the purposes of the planning system—an approach to candor which all should applaud. As this book goes to press, support for agriculture, housing, education, health and the poor is being reduced by a Republican administration because of a need to economize and because also of the low cost-effectiveness of these programs. Despite peace and the avowed need to economize, and with no claims for their cost-effectiveness, defense expenditures are being increased.

Thus the unequal development of the economy has, as its counterpart and mirror image, the unequal supply of public services. Those public services that are important for the planning system or that purchase its products are handsomely financed. Those services that are important for the market system or that have no industrial base—that, like relief of privation, provision of nontechnical education or administration of justice, serve the public at large—are far less amply provided. This distortion in priorities is not, as sometimes imagined, a sui generis error in an otherwise excellent system. It is as intrinsically a feature of the modern economy as erratic movement in an alcoholic.

The avowedly practical man should be as concerned with this characteristic of the economic system as the philosopher of self-confessed compassion. Government support is vital for the development of the economy—as a market for products, as a source of capital for encouraging and financing the development of new products and processes, for underwriting the risk that is associated with such development, for providing qualified manpower, for supplying highways, airways and other essential ancillaries of development, for rescuing defunct or unfortunate enterprises and for much else. No one gifted with willing sight fails to see the overdevelopment of the powerful industries. No less visible is the counterpart starvation of those industries, in the market system and the weaker end of the planning system, not similarly equipped to command public support. They do not have a guaranteed market in the state. Their products and processes get much less development assistance; their needs in manpower, ancillary public facilities, subsidy or underwriting of risk are not urgent public policy. Most important, capital which they might otherwise command goes to the part of the economy that is already overdeveloped. They must have recourse to overpressed and overcommitted lenders who charge accordingly.

Here, as earlier observed, is much of the explanation for the competitive decline of the United States in modern times. Numerous industries—textiles, shoes, railroads, shipping, machine tools—make or render old-fashioned products or services with obsolete equipment. The developmental energy and capital that might have altered this situation is invested in super-

sonic fighters, an antiballistic missile system and expeditions
to the moon, Mars and beyond.

2

The first need is to accommodate the division of revenue to
the division of tasks among the several levels of government.
The present arrangement was designed, one judges, with a
kind of malignant skill to serve not the public but the planning
system. Those services which are of the greatest importance
to the more powerful parts of the planning system are performed by the Federal Government. The personal and corporate income taxes which go to the Federal Government
expand more than in proportion with economic growth and
increasing income. The services to the planning system are
thus supported by automatically expanding revenues. The
taxes on which the states, and especially the cities, depend
have no similar upward flexibility. Sales taxes expand in rough
proportion to increases in income. The property tax on which
localities depend can only with the utmost effort be kept
abreast of increases in income.

Meanwhile the tendency of the civilian tasks of government is precisely the reverse of that of the revenues. With
the urbanization that follows from the decline in rural employment, increasing consumption and (even though the gain
is now more modest) increasing population, a larger share
of the tasks of government accrues to the cities. So it is with
the provision of housing, protection of persons and property,
provision of elementary education, protection of health, arrest
of air and water pollution, control of automobile use and
removal of the detritus of an increasing living standard. In
broad summary the revenues which expand with the economy
go to the Federal Government where they support the planning system. Those that do not so expand go to the cities
where they serve the public.

The preferred claim of the planning system on the more

ample revenues of the Federal Government is not an accident. Functions that serve the planning system—industrial research and development, support to technical education or the building of interstate highways are examples—come to be designated tasks of prime national importance. This reflects the interest of the planning system. Being so designated, they are appropriate functions of the Federal Government. A task that is historically, traditionally, logically or properly a function of state or local government is one that serves only the interest of the public at large.

There are, in principle, two remedies. One is to move public functions from the cities and the states to the Federal Government and thus allow them to participate in the more ample revenues of the latter. And the other remedy is to distribute some of the revenues of the Federal Government to states and cities. In practice both steps are required. Both are also part of the current debate. The effect of providing an alternative or guaranteed income (Chapter XXV) is to relieve states and cities of welfare costs at the expense of the Federal Government. So, in its fiscal impact, this also fits well into the general matrix of reform. It will be evident that revenue sharing, already forced in a limited way by circumstance, is directly a part of this reform. It is something to be pressed strongly and with particular emphasis on the claims of the cities. The public tasks of the large cities are more pressing than those of the states, and the revenue base on which the cities rely, the property tax in particular, is, as noted, markedly less elastic than the sources of revenue available to state governments.

3

The arrangement of functions within the Federal Government is also admirably accommodated to the needs of the planning system. The allocation of revenues as between services is all but exclusively the function of the Executive—the public bureaucracy. This is the branch of the government to which the planning system, as we have seen, has natural and effective access. And the effective power in this process rests with departments and agencies, the largest and most powerful of these being, predictably, the most effective in making its claims. The largest and most powerful of all is the Department of Defense, in relation to which the influence of the planning system is greatest. Charles L. Schultze, a former director of the Bureau of the Budget, has noted that the budget requests of the Department of Defense have, in the past, been largely immune to any basic challenge within the Executive. They served the higher purpose of security in the context of the Cold War. They were not questioned by the President; this being so, they could not be questioned by his subordinates.[1]

The protection against congressional intervention has been, and remains, if anything, even more complete. The examination of proposed public expenditures can be of two types: strategic and tactical. It can concern itself with the overall strategy of how revenues are to be distributed among functions, or it can seek to ascertain whether money for particular functions is needed or being well used—whether the allocation for paper clips, press agents or provisioning ships or diplomats is excessive. Even the decision as to the need for more destroyers or attack aircraft is tactical in character and only marginally affects the basic allocation as between functions.

Present congressional action on the budget is exclusively tactical. There is no consideration of the broad strategy of expenditure. Nor is there machinery for considering the allocation of resources among functions or occasion when this could be done. When it comes to the Congress, the budget is, in effect, split up between a large number of committees and subcommittees, each concerned with the legitimacy and need for the proposed outlays for the particular function. Thus

[1] "The Military Budget and National Economic Priorities," Hearings before the Subcommittee on Economy in Government of the Joint Economic Committee, 91st Congress, 1st Session, June 3, 1969, pp. 68, 72–73.

only tactical consideration is possible. Consideration of over-all priorities in expenditure, the vital matter, is effectively excluded.

It should be added that much of the tactical consideration of expenditures important to the planning system is perfunctory. The most important of these, notably those for defense, are protected by their technical character, secrecy, their great scale and the reliable servants of the Armed Services on the relevant committees. These legislators, as earlier noted, enjoy their power not as representatives of the public but as an extension of the public bureaucracy and the planning system. Public expenditures and those for the market system are more closely examined. No secrecy is there involved; they are more easily understood by the legislator; here legislators are more likely to make a point of their independence.

4

The President is decisively important for any remedy. On this, as on other matters, the all-important question in choosing a President must henceforth be whether the candidate distinguishes the planning from the public interest and is committed to the latter. But again the Congress is vital. It is the arm of the government that is least easily reached by the planning system. And it is the one most exposed to the pressures of public need and purpose—pressures that will grow as this purpose diverges increasingly from those of the planning system.

However, if the Congress is to be in any degree effective, it must be equipped for strategic as well as tactical consideration of the use of public resources. The basic and indispensable requirement is for a congressional budget committee.[2] This would receive the Executive budget and have plenary power

[2] Since this chapter was written, steps have been initiated looking toward some kind of congressional budgeting authority. Whether the results will be real or cosmetic, it is still too early to say.

to adjust its major categories in accordance with the public cognizance. Appropriations for defense, housing, welfare, education, agriculture, the cities would be thus considered in aggregate and in relation to each other. Minimal attention should be given to individual large items of expenditure within these categories and none at all to small ones. What is required is an exercise in planning—a consideration of the broad strategy of resource use in light of a similarly broad appraisal of public need. Needless to say this committee action should be well supported by staff. Since its concern is with distribution among competing functions, there should be no functional subcommittees. All needs should be pressed in relation to other needs.

Committee deliberations should include public hearings. These would provide occasion for full expression of individual and organized views as to the proper allocation of public resources. They would generate, inevitably, extensive public discussion. This would have an immediate effect in guiding congressional decision; it would have the more important consequence of making the allocation of resources a public issue. The attention so gained would be important for developing public concern for the issue—for crystallizing the public cognizance. It would also direct the attention of the electorate in salutary fashion to the position of individual legislators—to those who reflect the public cognizance as distinct from those who serve the planning system.

The allocation of resources settled upon in broad outline by the budget or planning committee would then go to the two houses of the Congress for debate, further adjustment and vote. The limits so established would then be binding on the legislative and appropriations committees and subcommittees. These latter would, as before, concern themselves pleasantly with the tactical employment of funds—with the need for, and legitimacy of, particular appropriations. They would normally be expected to appropriate the sums established by the budget committee. They could not, in the absence of further legislative action by the Congress as a whole, exceed the limits so established. Any amounts in excess of these limits would not be considered appropriated—would not be available for use. A certain amount of restructuring of the congressional committees might well be required so that committee action would fall within the broad planning categories.

5

The deliberations, hearings and debates on the allocation of public resources would be, as noted, the occasion for expressing the public cognizance—for pressing for the distribution of funds that expresses the public purpose as opposed to that of the planning system. For a long while certain simple presumptions can guide the budget committee and the resulting legislative action. It can be assumed that all appropriations which serve the planning system or purchase its products or services will be unduly generous. And it can be assumed that all appropriations for the public at large will be, relatively speaking, deficient. Resources should be reallocated accordingly.

There should also be a general presumption in favor of the needs of the less developed parts of the planning system and of the market system. Funds for research and development, manpower requirements and, most important, capital requirements for re-equipment and modernization in these parts of the economy will be inadequate. The deficiency will be greater with the diminishing power of the firm as one moves toward the market system. If there is to be a balanced development of the economic system—the kind of balance that even the most orthodox would hold to be necessary for efficient growth —reallocation is again necessary.

There should be a special presumption in favor of public support and assistance to the arts. As Chapter VII has shown, the planning system does not foster the arts—their resistance to organization places them outside its competence—and what the planning system does not need becomes inappropriate public policy. Highways for pleasure travel are of high public importance; museums and music for public enjoyment are of dubious public purpose. The need for special corrective action here will be evident.

Finally it can be assumed that the services rendered by cities and localities, and especially by the larger cities which are the product of especially rapid urbanization, will be underfinanced. Thus reallocation of revenue from the Federal Government to cities and localities will for a long while reflect the public cognizance.

6

All who are acquainted with the Congress of the United States —or with legislatures in general—will be aware of the difficulties of these reforms. The traditions and folk rites of the Congress greatly favor what exists over what is effective. Oratory and symbolic action are a substitute for serious change. Leaders now enjoy power and reward from serving the planning system or its symbiotic bureaucracies. It is their instinct that younger men who reflect the public cognizance had best be kept without power until they learn to get along by going along. Change, such as that here proposed, would enhance the power of the legislature but jeopardize the authority of those who now enjoy power.

Yet change is essential. It is a striking commentary on what is often called democracy in the United States that the most important issue before government—how the public moneys are to be used—is not now subject to legislative decision, is not even discussed. It is into this vacuum that the power of the planning system flows.

And, once again, there is the reinforcing power of circumstance. If the present system gave acceptable results, there would be no hope as well as little reason for change. But the unequal development of the economy is a fact and so is the contribution of government thereto. And so is the reciprocal distortion in government services. And these, in turn, are matters of major political discussion. Each new Congress in recent years has added new members giving new voice to dissatisfaction over the present distribution of public resources—to, in instinct if not in precise formulation, what is here called the public cognizance. The planning system has been effective and resourceful in defense of its favored position. (Currently it is being contended that public needs are not susceptible to being solved by money and—a less than novel thought—that public assistance on urgent problems suppresses private effort.) But, as ever, when persuasion is opposed to circumstance, it would be wrong to dismiss the force of circumstance. And as political instinct is reinforced by a clear view of the central reality of modern politics which is the conflict between the public and the planning purposes, the changes here urged will seem less remote. Or, since they are our salvation, so one must hope.

XXX

Fiscal Policy, Monetary Policy
and Controls

[He] showed remarkable courage in facing up to the problem
that this nation, and this Free Enterprise System, and this Free
Society faced. He enforced wage and price controls.
            —Former Secretary of the Treasury
            John B. Connally, speaking of President Nixon

THERE REMAINS THE QUESTION of how the planning and market systems are to be guided so that they yield a reliable flow
of income and product at reasonably stable prices. However
essential the other reforms they do not eliminate the need for
competent economic performance. One cannot have a socially
excellent economic system without having an economic system.
Fortunately equitable performance and effective operation of an economy go together. And failure is all but invariably the result of policies that are in the interest not of the
many but of the few, and invariably also with the pretense
that it is the many who are being served.

The usual reminder is in order on the problem to be solved.
The planning system cannot ensure that demand is sufficient to
keep the system operating at capacity. Decisions to save and invest are concentrated in a comparative handful of firms—a few
thousand. There is no machinery that ensures that the aggregate of the decisions to invest will be sufficient to offset the
aggregate of the decisions to save. If investment is insufficient,
the system will be subject to a downward spiral of output and

income which—by reducing investment more than saving—may for a substantial time be cumulative and persistent.

Nor has the planning system the power to arrest the counterpart upward spiral in prices. Nothing confines the modern union, in seeking pay increases, to what can be afforded at going prices. And the modern corporation has the power over prices that allows it to pass the resulting increases along to the public. Competition between unions and the need to anticipate increases in living costs from yet unrealized wage increases make the upward spiral also cumulative and persistent. Without action by the state the planning system is prone to depression or inflation.

Past and current action by the state to prevent depression or inflation has had five major shortcomings, to wit:

(1) The heart of the strategy for stabilization is a large public-sector expenditure supported by a progressive and flexible tax system. The public spending which is central to this process has been extensively in the service of the planning system. Its effect in distorting development, on income distribution, in depriving other parts of the economy of needed capital and in the potential for universal destruction has been sufficiently remarked.

(2) The supporting tax system has also come to reflect, increasingly, the preferences of the planning system—and notably the preferences of the higher-salaried members of the technostructure. In consequence it has become steadily less progressive in its incidence, steadily less responsive to increases and decreases in income and steadily less efficient for stabilizing income and expenditure.

(3) Further, in the last decade, when it has been necessary to increase demand, there has been general reliance on tax reduction instead of increased public spending. Economists and legislators, mistaking the planning for the public interest, have approved. Such reduction has extensively favored the higher incomes in the technostructure. Or, as in the case of the suspension of the automobile excise tax in 1971, it has favored

the products of the planning system. Or, as in the case of the tax credit for investment in the same year, it has directly subsidized the planning system. These tax reductions have increased the inequality in income distribution. They are also an inefficient method of expanding demand, for income is returned to affluent taxpayers where, in substantial part, it is saved. Savings from increased public expenditures to employ people on needed public tasks are much lower. Finally, when it is necessary again to restrain or contract demand, taxes are not easily increased. Instead exponents of the conventional wisdom call for economy in public outlays, and men of compassion are reduced to wishing they wouldn't. Since expenditures in behalf of the planning system are protected by pleas of high national purpose, it is spending for public purposes that is vulnerable.

(4) Those in charge of the management of the economy have been unable to perceive the decline of the market or—reflecting a natural vested interest in painfully acquired knowledge—have been unwilling to concede it. Accordingly faith in orthodox fiscal and monetary policy has died slowly, and efforts to come to grips with the wage-price spiral have been apologetic, halfhearted and avowedly temporary.

(5) Finally the efforts both to expand and contract demand have invoked monetary policy. This operates with punishing difference in effect as between the market and the planning systems.

2

The remedy begins not with the needs of stabilization but with those of general reform. And reform also requires a large and stable flow of public expenditures—expenditures that are related to public purposes, not those of the planning system. And an equitable distribution of income requires that these be paid for with a strongly progressive tax structure—a tax structure that reflects the public interest in fairness, not the interest of the planning system and the constituent technostructures in their own reward. Both these expenditures and these taxes accord exactly with the needs of stabilization policy.

For supporting the economy, expenditures that reflect the public cognizance are categorically more efficient than those that serve the interests of the planning system. The latter go in large amounts for higher-salaried members of the technostructure or for profits. In both cases the ratio of saving to income is high. Such saving does not add to demand. Spending for public purposes, by contrast, is, in greater part, for ordinary salaries or wages or it is in the form of payments for pensions, unemployment compensation, income maintenance or other assistance to those in need. Saving is much smaller and from some classes of expenditure nonexistent. So, much more of what is spent adds to demand. In the United States, it should be kept in mind, there is no net saving among people in the lower half of the income scale.

Similarly with taxation. Those taxes that serve the goal of greater equality are those that are most efficient for stabilization. The corporation and personal income taxes do most to equalize income. They are also the taxes that increase more than proportionately with increased income and purchasing power and decrease more than proportionately with decreased income and purchasing power. Thus they are the taxes that serve best the goals of stabilization. And the more substantial (within all reasonable limits) the reliance on the corporation income tax and the more progressive the personal income tax, the greater both the stabilization and the equalization effect.

The more comprehensive the tax system—the fewer the loopholes—the better it serves both equality and stabilization. The avowed purpose of special concessions or loopholes is to stimulate particular classes of economic activity. In practice the beneficiaries are far more often firms or individuals in the planning system than those in the market system.[1] The stimu-

[1] The result, in particular, of the preferential taxation of enrichment in the form of corporate capital gains.

lation, so far as there is any, is to the overdeveloped not the underdeveloped part of the economy. The proper rule in the modern economy is to treat all enrichment alike—to apply a common rate of taxation to any enrichment whether it accrues in the form of salary, capital gains, property income, inheritance, gift or, since one must be meticulous, larceny, fraud or embezzlement. The enrichment is the basic fact; given that, the tax follows. This rate, since there would be no un-taxed income, could be lower than under the present system of selective taxation of income. It would, of course, be strongly progressive.

In recent years the discussion of taxation has been increasingly along the foregoing lines. A Canadian Royal Commission has proposed such a tax system; Joseph A. Pechman, perhaps the leading American tax authority, has made similar proposals. It is a reform of great and urgent importance both for the more effective operation of the modern economy and for its civilized effect on income distribution.

3

To summarize, the proper fiscal policy begins with the level of public expenditure. This is not given by the needs of fiscal policy itself; it is given by the need for public services as opposed to those supplied by the market and planning systems. The level of public expenditure so established gives, in turn, the amount of taxation required.

There is no assurance that the combination of expenditure and progressive taxation just given will yield the proper level of demand. It may be too great or too small by the tests presently to be mentioned. If demand is excessive, the *generally* appropriate procedure will be to increase taxes. The required level of public outlays has been decided on the basis of need. That the planning system uses its power to win priority for the private consumption of its products has been sufficiently established. This includes priority over public services not important to itself. The effect of tax increases is to cut back on less important private consumption and to protect more important public consumption. To the extent that the tax system is progressive and the tax increase falls more heavily on the affluent, the case for reducing private as opposed to public consumption is that much stronger.

If demand is deficient, the generally proper procedure will be to increase public expenditure. This, as earlier noted, is the most efficient way of adding to demand; it also reflects the generally higher need for public, as opposed to private, consumption.

It will be suggested—and by some trumpeted—that the policy here proposed means, over time, an upward drift in taxation. This is so. But this only means that fiscal policy would correct the general bias of the economic system in favor of the products of the planning system—the products that reflect, among other things, its superior powers of persuasion. And no policy is forever. Conceivably the time may come when public needs —including those of the large cities—are as amply supplied as the present private consumption of those who pay income and corporation taxes. When that day comes, it will be sufficiently noticed and celebrated. Then it will be time to use tax reduction as a corrective for a deficiency in demand.[2]

4

The next step in reform is to reduce, and for all time, the use of monetary policy. Such policy works by reducing or increasing, directly or indirectly, the amount of money available for

[2] There is a case for a type of public expenditure that increases more or less automatically with an increase in unemployment and reduces itself as unemployment diminishes. This is so-called public service employment—the use of the state as a general employer of last resort. Resulting employment would be auxiliary to present sanitation, park, police, custodial, health and primary, secondary and higher education employment. The administrative problems— including the relation of their levels of compensation to those of regularly employed workers—are considerable. So, however, are the advantages to the people who get the work and the community which gets the services.

lending. Those who least need to borrow and those who are most favored as borrowers are in the planning system. Those who most rely on borrowed funds or are least favored at the banks are in the market system. The planning system is the most highly developed part of the economy, the market system the least developed. Monetary policy thus favors the strongest and most developed part of the economy, discriminates against the weakest and least developed part.

Monetary policy is also, as a technical matter, highly uncertain in its effects. No one knows what the response to a greater or less availability of funds for borrowing will be or when that response will occur, for the reason that the factors that govern such response are never the same from one time to the next. This uncertainty is concealed, in turn, by intense and solemn discussion employing arcane terminology—rediscount rate, prime rate, open market purchases, spreads, twists—under conditions of priestly seclusion. The public wrongly assumes that this discussion proceeds from knowledge. In fact where there is knowledge and certainty—when men know what will happen as a result of a given action—there is little to discuss.

The interest rate, like other prices in the planning system, is now fixed—for practical purposes by the Federal Reserve System as it influences the rates charged by the banks. The first step toward a proper policy consists in accepting the notion of a relatively permanent level and structure of interest rates. It would no longer be active policy to limit borrowing—and therewith the volume of spending from borrowed funds and therewith, also, for that matter the supply of money—by raising interest rates or by restricting the supply of loanable funds available at the going rate, this being the first step toward higher rates.

The level at which the interest rate should be set is largely arbitrary—but the bias should be on the low side. Low interest rates favor borrowers rather than lenders. As a broad and not excessively astonishing rule, borrowers have less money than lenders. So lower interest rates contribute to a more equitable distribution of income. And since it is the market system that depends most on borrowed funds, low interest rates favor the development of this part of the economy which also accords with need and public purpose.

Although in a proper policy the interest rate ceases to be the instrument for controlling the volume of borrowing, the volume of borrowing *does not* remain uncontrolled. Control is exercised through tax and expenditure policy on the volume of demand in the economy. Borrowing will be excessive when, along with other sources of demand, it is pulling up prices. It is then curtailed by increasing taxes; this reduces the capacity of people to incur mortgage or other personal debt, and, as the demand for goods recedes, it reduces both the incentive and the ability to borrow or use funds for business expansion. The taxes which serve this policy strike large and small firms alike and rich taxpayers as well as poor. And if the taxes are progressive, as just urged, they hit the most affluent the most. This policy is equitable in its application. And since it does not depend on intrinsically unpredictable responses to changes in the interest rate or the supply of loanable funds, it is far more certain in its effects.[3]

In economics there are few absolutes. There may be occasions when a general excess of demand derived from excessive borrowing justifies an increase in interest rates. And there may be occasions when borrowing can be encouraged with lower rates. But it must be a prime tenet of effective and equitable economic management that such changes be exceptional. Any active monetary policy operates by recurrent and discrimina-

[3] A similar if slightly less severe recommendation for minimizing the use of monetary policy is in Arthur Okun's "Rules and Roles for Fiscal and Monetary Policy" in *Issues in Fiscal and Monetary Policy: The Eclectic Economist Views the Controversy,* edited by James J. Diamond (Chicago: DePaul University, 1971). The first rule of stabilization policy he suggests is to *"Keep monetary conditions close to the middle of the road,"* and the second is to *"Operate fiscal policy to avoid forcing monetary policy off the middle of the road."* The case for a strong fiscal policy based on public service employment has been effectively advanced by Melville J. Ulmer, "Toward Public Employment and Economic Stability," *The Journal of Economic Issues,* Vol. VI, No. 4 (December 1972), p. 149.

tory reduction in investment in the weakest part of the economic system. (The case of housing is especially dramatic.) It thus contributes directly to inequality in income and inequality in development. It thus intensifies the central and most painful faults of the modern economy. And, all but sadistically, it puts the pain on those least able to bear it.

The service of economics to special interest is no new subject, but the subtlety of the service where monetary policy is involved cannot be too much admired. Although the discrimination is palpable, nearly all of the discussion holds it to be socially neutral. And the priestly character of the discussion disqualifies the potential critic—and causes him, indeed, to disqualify himself. In the past a vague populist instinct has identified monetary policy as socially regressive and brought its instruments, notably the Federal Reserve System, under a measure of attack. But even the most liberal economists have been at pains to dissociate themselves from a current of criticism which all reputable men were required to regard as unlearned and naïve. As so often what was reputable was what served the powerful and the affluent.

5

The oldest goal of monetary and fiscal policy, one that comes automatically to the tongue of the most primitive scholar, is full employment at reasonably stable prices. We must now leave this ancient intellectual anchorage; truly nothing in life is ever quite secure. Full employment, we have seen, means wages and working conditions for numerous workers that are socially intolerable. Lack of qualification and unsuitable location (both, in part, a disability of race) commit many people to the market system. There they can find employment only by accepting derogatory work at low wages or by self-exploitation. The purpose of the alternative or guaranteed income is to provide a decent alternative to such work and pay. It accepts that some work and pay are worse than unemployment.

The modern test of fiscal policy is a level of output in the planning system which employs at a decent wage the supply of available and qualified workers, allowing always for those who are changing jobs or whose skills have become obsolete. It is not a test of success that all workers everywhere be employed. For those whose qualification and location do not allow of employment at a socially decent wage, resort to a guaranteed or alternative income must be accepted.

The test of policy for the planning system is the level of output in relation to the supply of available and qualified workers. For the market system the test is the behavior of prices. Given a proper policy, prices in the market system will be generally stable except as there may be increases resulting from an improved bargaining position, higher minimum wages or other steps toward greater equality in relation to the planning system.

An excess of qualified workers seeking employment in the planning system, together with falling prices (or even stable prices) in the market system, will suggest a deficiency of demand that should be rectified. A shortage of qualified workers in the planning system and a persistent upward thrust of prices in the market system are an indication that demand is excessive and should be curbed.

Fiscal policy remains essential for maintaining a general balance between demand and supply. No modern development makes this policy—or the resulting balance between aggregate demand and supply—less important. This is a delicate matter. To a point strong demand improves the terms of trade of the market system. But inflation here—a matter of recent experience as this is written—signals also the need for more astringent fiscal policy.

6

The final step in the general management of the economic system is the regulation of wages and prices in the planning system. Here the market has been disestablished; the requisite planning to maintain price stability is beyond the competence of the individual firm. In the absence of action by the state there is, in consequence, a steady and accelerating upward spiral of wages and prices. Public control is thus inevitable.

Price and wage controls, as we have seen, are psychologically the most difficult of all the measures here proposed for the defender of the established view to accept. Other instruments of planning employed by the planning system—control of individual prices, control of costs, organization of supply at those costs, provision of an internal source of capital, persuasion of the state as regards needs, management of the state as regards procurement—can be ignored or minimized by the person who is determined to do so. It takes effort to achieve this absence of perception, but the rewards in intellectual and pecuniary capital conserved are great. The market survives. With wage and price controls the game is up. A market system in which wages and prices are set by the state is a market system no more. Only the blithely obtuse can reconcile "this Free Enterprise System" with the enforcement of wage and price controls.

This psychological barrier is, in turn, a factor of first importance in the management of controls. Such management tends to be by men who find such action inconsistent with their deeper faith. It is the problem of the abortionist who is a devout Catholic, the committed voyeur who leads the pornography squad. No one suggests that the unions and the corporations that make the controls inevitable are a temporary phenomenon. But the faith remains strong that what one does not wish to exist need not exist. God rules, and God cooperates with good conservatives. If there must be a secular reason, it can be invented.[4,5]

---

[4] Thus the following in the 1972 Economic Report of the President: "The basic premise of the price-wage control system is that the inflation of 1970 and 1971 was the result of expectations, contracts, and patterns of behavior built up during the earlier period, beginning in 1965, when there was an inflationary excess of demand. Since there is no longer an excess of demand,

The first requirement is that controls be seen as forever—or at least for as long as there are unions and corporations in their present relationship in the planning system. Without such recognition the policy will be like a pendulum: Inflation or unemployment or both being unacceptable, there will be pressure for controls. These will be adopted and continued until they appear to be working. Then they will be abandoned, and the movement will start anew. Meanwhile little or no intelligence or energy will be brought to bear on design or administration —on making them workable and equitable.

Given the acceptance of controls, five basic rules govern their administration. The rules, not surprisingly, follow from the argument of the foregoing chapters. They are:

(1) The controls need apply only to wages that are set by collective bargaining and to the prices of firms that are in the planning system. This means that, in the United States, price controls need not apply to more than a few thousand of the largest firms. In the market system stability or approximate stability is ensured not by price control but by fiscal policy. And in the planning system, while the wage and price controls prevent the price increases that are the result of the interaction of wages and prices, demand must not be in excess of what can be supplied at current prices. This eliminates any tendency to sell above the established price—for a gray or black market price to develop—by the remarkably effective device of having all that is wanted available at the legal price. Controls, to repeat, are not a substitute for a fiscal policy that ensures a broad equivalence between aggregate demand and what the economy

the rate of inflation will subside permanently when this residue of the previous excess is removed. The purpose of the control system is to give the country a period of enforced stability in which expectations, contracts, and behavior will become adapted to the fact that rapid inflation is no longer the prospective condition of American life. When that happens controls can be eliminated." Economic Report of the President, 1972, p. 108.

[5] Thomas Balogh, the pioneer advocate of the control policies to which the British government has reluctantly come, argues that no one who transgresses is ever quite forgiven. Better a true defender of the faith than a man who is right by being wrong.

can supply. They are only an essential supplement to such equivalence.

(2) Controls need not freeze either wages or prices. The essential bargain, one that now has a measure of acceptance in principle, is that wage increases should be confined to average productivity gains in the planning system. Wage costs being thus constant, prices can remain generally constant. However wage increases cannot be denied to unions in industries where productivity gains are less than average. Because their particular industry does not have productivity gains cannot be a ground for discriminating against those there employed. And compensating price increases cannot be denied to firms in those industries.

Nor is it generally necessary to fix the prices of the individual products of a firm. An instruction against increasing the weighted average of its prices in a given product line will usually be sufficient. While large retailers may be enjoined against widening their margins, retail price control is not a central necessity. A substantial part of retailing is in the market system, and retailers, in general, do not possess great market power.

(3) Wage controls should not freeze wage or income differentials. On the contrary there should be affirmative effort to narrow such differentials. To accept controls, we have seen, is to abandon the pretense that compensation is established by the market. It is the product of human agency; power is decisive in determining who gets how much. This being recognized, the public concern for a more equitable distribution of income requires that the results of such exercise of power be minimized.

Within the corporation this means that wage increases resulting from productivity gains should go in greatest amount to those who get the least. There should be a strong and positive effort to narrow the present gap between worker and executive. At a minimum the allowable increases should decrease (and ultimately to zero) at higher levels of pay.

Within the planning system lower-wage industries should be allowed to catch up with higher-wage industries. Workers in the market system, some exceptional cases involving small employers but strong unions apart, should not be subject to controls.

(4) It follows from the foregoing that, however successful the system of controls, prices will not be stable. Inequality and inequity in the application of the controls, as well as hardship, should all be mitigated by an evening-up process—by allowing those prices or wages that are too low to rise. Ordering the reduction of the higher prices is rarely if ever practical. So there will be a continuing upward drift in prices. It is important that the increases so occasioned—increases that are the result of equalizing adjustments *within* the system—be distinguished from those that result from the crude and general interaction of all wages on all prices in the planning system. It is the second, not the first, that the controls are designed to prevent.

(5) Finally, if the controls are to be effective, there must be the requisite exercise of public power. The problem here needs to be seen with far greater clarity than in the past. When prices and wages are established by the market, it is believed—or, in any case, the relevant theory holds—that the ultimate decision is that of the public. By its decisions to buy or not to buy and to buy what, the public instructs the market, and the market establishes the level of prices and the resulting compensation. It follows that, if the government interferes with this process by fixing wages and prices, it interferes (it may be supposed arbitrarily) with what is already a public decision.

But controls are made necessary because planning has replaced the market system. This is to say that the firm and the union have assumed the decisive power in setting prices and wages. This means that the decision no longer lies with the market and thus with the public. It lies with the planning system. Intervention by the government is not in a public process. Intervention is in a private decision that is in pursuit of private goals. Government intervention, if it reflects the public cognizance, means public government instead of private government.

Thus the justification for the exercise of public authority. And effective exercise of such public power is essential. In the setting of wages and prices there should be extensive consultation with industrial firms and unions. This is especially so in the case of the unions. But in the end the government must ensure compliance with the goals specified above. There has long been a belief that the fact of control can somehow be obscured by keeping compliance on a voluntary basis. This is a delusion: The market is no less dead because the controls that replace it are unenforced. And the effect of voluntary control is primarily to reward those who are least inclined to comply. Needless to say, also, there can be no control which merely ratifies what corporations and unions agree on anyway—which sanctifies the result that would be reached without any control.

7

In the administration of price and wage controls, as in lesser measure in the guidance of fiscal policy, there is a dilemma: A major role must be assumed by the Executive, and the specific tasks involved could hardly be more exposed to the resulting influence of the planning system. Yet these are tasks which, in their performance, must reflect the public cognizance.

There is no easy answer. The only hope is a publicly cognizant President and, above all, a publicly cognizant and vigilant legislature. Fortunately the rules that reflect the public interest are rather simple. If public expenditures are increasingly for public purposes, if taxes are increasingly progressive, if monetary policy is passive, if expansion of demand is accomplished by increased public expenditure and contraction of demand by increased taxes, if wage increases are kept in accordance with productivity gains, if increased equality is a major consideration in making wage adjustments and if price increases are allowed only in response to hardship resulting from the evening-up of wages and the absence of productivity gains—then an essentially public management is being achieved. The enforcement of such rules is not beyond the capacity of a publicly cognizant President and legislature. What is suggested also accords with progressive instinct in these matters. But energy and vigilance are certainly required.

XXXI

Coordination, Planning
and the Prospect

IN 1945, when World War II came to an end, the railroads of the United States were carrying a record volume of both freight and passengers. They were not, however, a particularly powerful part of the planning system. Tradition, regulation and the relatively nontechnical character of the enterprise had kept the railroads from developing a strong technostructure and therewith a powerful position in relation to their customers and the community, and in the state. In the 1950s, the far more powerful road transport industry, led by the automobile companies, initiated the new interstate highway system. Rail passenger travel gave way to the passenger car with a lesser loss to the also influential (and strongly supported) airlines. And rail freight traffic was diverted extensively to the highways.

The organization by the automotive industry of the increased use of its products was, as the preceding pages have stressed, a triumph of the planning system. The same development, accentuated by parallel changes in the electric utility and home-heating industries, also greatly increased the demand for petroleum products. The full magnitude of the increase in demand was not foreseen. Nor was it foreseen that the building of the needed pipelines, refineries and docking and discharging facilities for large tankers would involve difficult and, on occasion, irreconcilable environmental conflicts. In consequence there is now doubt as to whether the petroleum products made necessary by the great expansion in automotive, truck and other use will be available. A new phrase has been born. There is much talk of an "energy crisis."

As with other faults of the planning system the possible shortage of petroleum products is being discussed as though it were sui generis—another purely adventitious failure. We now know that it is nothing of the kind. The planning system involves an intricate coordination as between its several parts in pursuit of their purposes. There is every likelihood that, from time to time, this coordination will fail. And these failures are already fairly common. There is, strictly, no electricity shortage. Rather the promotion of electrical use by appliance manufacturers—including the promotion of architecture for which air conditioning is essential—is running ahead of the ability of utilities to supply electricity.

When, as in the case of the automobile and the oil industries, planning by one industry imposes requirements on another that it cannot meet, it is, of course, taken for granted that the state will intervene. By public action freight and people will be returned to the more economically fueled railroads. Or there will be subsidy to hitherto unprofitable sources of energy. Or there will be technical support to the development of new energy supplies. The state, in short, will take steps to effect the coordination of which the planning system is incapable. It will impose overall planning on the planning system. This is the next and wholly certain step in economic development—one that is solidly supported by the logic of the planning system. Only with difficulty will it be described as a way of strengthening the market and the free enterprise system.

The solution is to recognize the logic of planning with its resulting imperative of coordination. And government machinery must then be established to anticipate disparity and to ensure that growth in different parts of the economy is compatible. The latter on frequent occasion will require conservation—measures to reduce or eliminate the socially least urgent use. On other occasion it will require public steps to expand output. The sooner the need for such action is recognized, the less the inconvenience and suffering from the crises that are now predictable and for which there is no other remedy.

There will have to be a public planning authority. This, in turn, will have to be under the closest legislative supervision. For here will be encountered the most difficult of all the problems of the public cognizance. That will be to have planning that reflects not the planning but the public purpose. The creation of the planning machinery, which the present structure of the economy makes imperative, is the next major task in economic design.

2

A second major problem of coordination of planning systems is on the horizon—or, more precisely, well up on this side. It involves the systems of different countries. Again the foregoing chapters suggest the shape of things.

The firms that comprise the planning system, we have seen, stand astride national frontiers. They largely dispense with tariffs as a hindrance. They take their products into other industrial countries where they join the oligopolistic convention that establishes prices there. They expand production—and therewith investment—in those countries where the cost is lowest. Those firms that are most strategically situated in the countries of lowest-cost production—normally those that have their headquarters in such countries—expand the most.

The cost advantages just mentioned are of three kinds. There is the classical possibility that the labor force works more for the same pay, as much for less or more for less. Or the capital equipment may be cheaper or technically better—more modern. Or the country has had a lower rate of inflation. The wage-price spiral has been under better control or, for other reasons, works with less punishing effect.

In recent times the foregoing advantages have strongly favored Japan and Germany. Both have a diligent labor force, and that of Japan is at lower cost. In the United States, as earlier noted, the symbiotic relationship between the planning system and the public bureaucracy has led to heavy investment in industries, notably weapons production and space exploration, to the neglect of civilian industry for which capital has recurrently been expensive and scarce. Germany and Japan have, by contrast, had abundant funds for modernizing and expanding less powerful civilian industry. And, until comparatively recent times, the demands of German and Japanese unions have been less vigorously pressed than those of the unions in the United States and Britain.

The migration of production to the advantaged planning systems means that these countries—more precisely firms in these countries—accumulate balances in the currency of the disadvantaged country to which they are selling and from which they are buying much less. Recurrently these balances inspire uneasiness in their owners; an effort is made to convert them into the currency of the advantaged country. This encounters natural reluctance. Why exchange a strong currency, one that has every chance of appreciating in relative value, for one for which the prospect is the reverse? The consequence of this effort and this reluctance is the most familiar phenomenon in the relationship between the planning systems of different countries. It is known as the currency crisis. In all recent manifestations the dollar has been the crisis currency. It has been in Japan and Germany that the dollars have accumulated, and it is into their currencies, and one or two others, that conversion is sought.

Adding to the confusion of the currency crisis is the discussion of it. This is partly fraudulent, partly incompetent and for the rest extensively irrelevant.[1]

[1] The layman may think these hard words—the agreeable intraprofessional calumny in which scholars regularly rejoice. Alas, they are measured and true—as a moment's reflection will establish. We have been having currency

The problem of currency relationships seems impenetrable to the ordinary or even the learned citizen. In this situation the currency expert, whose ignorance of that of which he speaks is frequently concealed even from himself, flourishes. Incompetence is inherent in the curious convention that holds that any individual, however inadequate or obtuse, on becoming a Secretary of the Treasury, an Undersecretary for Monetary Affairs or a member of the Federal Reserve Board in the United States, or a counterpart official abroad, becomes, by virtue of desk and office, fully qualified in the subject.

Irrelevance is much more serious; it follows, not surprisingly, from the familiar commitment to the market and the neoclassical faith. Given this, the problem of coordination can be resolved in the short run by the devaluation of the currency of the disadvantaged country which makes its products cheap in foreign currencies and countries and makes foreign products more expensive in its own markets. Then, with a little more time, a standard monetary and fiscal policy will bring inflation to an end, if that is the problem. And, given a little more time, capital will flow to the laggard industries in the disadvantaged country with corrective effect, and diligent workers in the advantaged countries will demand their due in higher wages with higher costs and prices and further corrective effect. From all this comes the belief that there *is* a monetary solution for trade and currency problems as between industrial countries. One need only be an expert and find it.

When the problem is between planning systems, all of the foregoing is readily seen to be a mirage. The prices of the decisive manufactured products of the advantaged countries in the disadvantaged one—those of Germany and Japan in the United States, for example—are part of the oligopolistic convention

---

crises for years. For years the experts have been meeting, with all proper solemnity, on a solution. Had they a solution, it is inconceivable that it would not, before now, have been adopted. It is in the nature of a solution that the problem is solved. Had the problem been solved, there would be no currency crises. They continue. Also we have come to accept that, while monetary experts will meet and discuss, they will not solve problems.

of the receiving country. Devaluation does not automatically increase prices; it is possible that the firms in the supplying country will reduce prices, accept lower margins and revenues and continue the volume as before. And, if that is not possible, they will act with all vigor to have their governments resist the devaluation by the disadvantaged country. This can easily be accomplished by devaluing too.[2] The planning systems of these countries have the power common to such systems in relation to the state to be persuasive on the point. For its primary effect devaluation depends on the products (and services) of the market system that enter international trade.

And further there is no longer a tendency for things to work out. Advantage and disadvantage are associated with the distribution of capital as between industries. There is no tendency for countries, such as the United States, which have a bad distribution of capital based on a symbiotic association between weapons industries and the public bureaucracy, to correct matters. Inflation is the result of the power of corporations and unions and the lack of efficacy of the controls. Differences here are not ironed out by adherence to a common set of monetary and fiscal precepts.

The only remedy is in the coordination of planning policies as between the national planning systems. This must include common policies in the distribution of capital as between industries, common steps to control the wage-price spiral. In the absence of public authority coordinate with the international scope of the task, the difficulties are obvious. They would be less were the planning tasks of the United States fully perceived and effectively carried out, leaving other countries the problem of adjusting their planning thereto. In the years following World War II, the international system worked because American policy was reasonably predictable and the smaller countries adjusted their policies to that of the large one. Until this arrangement is restored, one thing is certain: The planning

[2] Or by resisting revaluation, which is the modern remedy pressed on the advantaged country.

systems of the several industrial countries will continue, as in the recent past, to stumble from one so-called monetary crisis to the next. The monetary experts will travel, convene and confer in the secure knowledge that nothing they do will make their movement or their employment redundant. After recrimination based on differences that people will not understand, devaluation or revaluation will be agreed upon. This will be hailed as a solution, and the next crisis will then be in the offing. Eventually, perhaps, the lesson will be learned. National planning systems, operating internationally, also require a measure of international planning. It is not a subtle point.

3

The time has come for a concluding word on economics. Lord Keynes, in a famous forecast, thought the subject would eventually become unimportant—in social significance it would rank about with dentistry. Not everything that has here been said about economics has been kind, although few who give the matter thought will believe the present strictures unjust. Economics is a vast profession; much is spent on its research and pedagogy. If all were well with the subject, we would not be suffering from so many unsolved and unexpected problems. But, though in a sense Keynes was right that the subject is in decline, in a larger sense he was wrong.

He was right to the extent that economics is concerned with the production of goods and the prevention of depressions. In the modern industrial society these are not very difficult matters. Those who are concerned with them may be socially more important than those who relieve toothache and remove abscessed teeth, but not much. In trying to place all problems within the framework of the market and all behavior subordinate to market command, economists do, we have sufficiently seen, render great service to the planning system—to the disguise of the power that in fact it wields. But this is socially a dubious function and not one that we need applaud.

But in a more acceptable sense Keynes was wrong. He did not see that, with economic development, power would pass from the consumer to the producer. And, not seeing this, he did not see the increasing divergence between producer or planning purpose and the purpose of the public. And he did not see that—since power to pursue the planning purpose is unequally distributed—development would be unequal. And therewith the distribution of income. Nor did he see that the pursuit of such purpose would threaten the environment and victimize the consumer. And he did not see that the power which allows producer purpose to diverge from public purpose would ensure that inflation would not yield to a simple reversal of the policies that he urged for unemployment and depression. Nor did he foresee the problems of planning co-ordination, national and international, just mentioned.

Given what was not foreseen, the future of economics could be rather bright. It could be in touch with the gravest problems of our time. Whether this is so—whether economics is important—is up to the economists. They can, if they are determined, be unimportant; they can, if they prefer a comfortable home life and regular hours, continue to make a living out of the infinitely interesting gadgetry of disguise. They will, as was the case in the summer of 1971 when price controls were imposed in the United States, or a year later when they were put into effect in Britain, have few or no words of guidance or advice on great issues. They will be socially more irrelevant than Keynes's dentist, for he would feel obliged to have a recommendation were everyone's teeth, in conflict with all expectation, suddenly to fall out.

Or economists can enlarge their system. They can have it embrace, in all its diverse manifestations, the power they now disguise. In this case, as we have seen, the problems of the world will be part of their system. Their domestic life will be less passive. There may be a contentious reaction from those whose power is now revealed and examined. And similarly from those who have found more comfort than they knew in the fact that economists teach and discuss the wrong problems or none at all. But for a very long while to come economists will thereby escape the fate that Keynes foresaw.

INDEX

Index